ANIMAL BIOTECHNOLOGY

Science-Based Concerns

Committee on Defining Science-Based Concerns Associated with
Products of Animal Biotechnology

Committee on Agricultural Biotechnology, Health, and the Environment
Board on Agriculture and Natural Resources
Board on Life Sciences

Division on Earth and Life Studies

NATIONAL RESEARCH COUNCIL
OF THE NATIONAL ACADEMIES

THE NATIONAL ACADEMIES PRESS
Washington, D.C.
www.nap.edu

THE NATIONAL ACADEMIES PRESS   500 Fifth Street, N.W., Washington, DC 20001

NOTICE: The project that is the subject of this report was approved by the Governing Board of the National Research Council, whose members are drawn from the councils of the National Academy of Sciences, the National Academy of Engineering, and the Institute of Medicine.   The members of the committee responsible for the report were chosen for their special competences and with regard for appropriate balance.

This study was supported by Contract No. 223-93-1025 between the National Academy of Sciences and the U.S. Food and Drug Administration, and funds by the National Research Council.   Any opinions, findings, conclusions, or recommendations expressed in this publication are those of the author(s) and do not necessarily reflect the views of the organizations or agencies that provided support for the project.

This report has been reviewed by a group other than the authors according to procedures approved by a Report Review Committee consisting of members of the National Academy of Sciences, the National Academy of Engineering, and the Institute of Medicine.

Library of Congress Cataloging-in-Publication Data

Animal biotechnology: science-based concerns / Committee on Defining Science-based Concerns Associated with Products of Animal Biotechnology, Committee on Agricultural Biotechnology, Health, and the Environment, Board on Agriculture and Natural Resources, Board on Life Sciences, Division on Earth and Life Studies.
      p. cm.
Includes bibliographical references and index.
  ISBN 0-309-08439-3 (pbk.)
1. Animal biotechnology. I. National Research Council (U.S.).
Committee on Defining Science-based Concerns Associated with Products of Animal Biotechnology.
  SF140.B54 A58 2002
  660'.65—dc21
                              2002151075

Additional copies of this report are available from the National Academies Press, 500 Fifth Street, N.W., Lockbox 285, Washington, DC 20055; (800) 624-6242 or (202) 334-3313 (in the Washington metropolitan area); Internet, http://www.nap.edu-

Printed in the United States of America

THE NATIONAL ACADEMIES
Advisers to the Nation on Science, Engineering, and Medicine

The **National Academy of Sciences** is a private, nonprofit, self-perpetuating society of distinguished scholars engaged in scientific and engineering research, dedicated to the furtherance of science and technology and to their use for the general welfare. Upon the authority of the charter granted to it by the Congress in 1863, the Academy has a mandate that requires it to advise the federal government on scientific and technical matters. Dr. Bruce M. Alberts is president of the National Academy of Sciences.

The **National Academy of Engineering** was established in 1964, under the charter of the National Academy of Sciences, as a parallel organization of outstanding engineers. It is autonomous in its administration and in the selection of its members, sharing with the National Academy of Sciences the responsibility for advising the federal government. The National Academy of Engineering also sponsors engineering programs aimed at meeting national needs, encourages education and research, and recognizes the superior achievements of engineers. Dr. Wm. A. Wulf is president of the National Academy of Engineering.

The **Institute of Medicine** was established in 1970 by the National Academy of Sciences to secure the services of eminent members of appropriate professions in the examination of policy matters pertaining to the health of the public. The Institute acts under the responsibility given to the National Academy of Sciences by its congressional charter to be an adviser to the federal government and, upon its own initiative, to identify issues of medical care, research, and education. Dr. Harvey V. Fineberg is president of the Institute of Medicine.

The **National Research Council** was organized by the National Academy of Sciences in 1916 to associate the broad community of science and technology with the Academy's purposes of furthering knowledge and advising the federal government. Functioning in accordance with general policies determined by the Academy, the Council has become the principal operating agency of both the National Academy of Sciences and the National Academy of Engineering in providing services to the government, the public, and the scientific and engineering communities. The Council is administered jointly by both Academies and the Institute of Medicine. Dr. Bruce M. Alberts and Dr. Wm. A. Wulf are chair and vice chair, respectively, of the National Research Council

www.national-academies.org

v

Committee on Agricultural Biotechnology, Health, and the Environment

Board on Agriculture and Natural Resources

Preface

What we have before us are some breathtaking opportunities disguised as insoluble problems.

—John W. Gardner, 1965, upon appointment as the Secretary of the Department of Health, Education, and Welfare

Rarely in the modern history of humans has biology played such an important role in human affairs as it does today. In the eighteenth and nineteenth centuries, explorers stimulated the first major advance in biology by bringing back countless new species that Darwin, and others, put into a logical order based on the theory of natural selection. The development of evolutionary thinking and the clarification of the rules of genetic inheritance resulted in the theoretical base for targeted artificial selection—an essential component of progress in biology and agriculture.

A second major advance currently is underway. Due to the basic understanding of inheritance at the molecular level and the tools this has made available to biologists, it no longer is necessary to depend upon natural or artificial selection and breeding of progeny to produce new and improved individuals. Genes from the same or other species can be inserted into a genome, or the activity of a specific gene can be blocked. Further, once the genome has been altered artificially, large numbers of new plants and animals carrying the modified genome can be made using cloning techniques. Producing animal models of human diseases for research, improving medical procedures, and increasing food production are but three modern advances that already have come to pass. More advances are predicted for the future. The committee—early in its discussions—recognized that not everything that bloomed from the biotechnology garden was a flower ready to be picked for the human bouquet. As was true for other technologic advances in the past, advances do not come without expected and unexpected risks. The committee also recognized that the technology it was studying is in its infancy. Many of the problems, such as inefficient reproduction and production of abnormal offspring, are receding as the technology advances. Therefore, the committee presents a "snapshot" of biotechnology and of potential concerns about that technology at present.

In view of the rapidly-changing biotechnologic landscape, federal agencies with responsibility for ensuring food safety, maintaining modern medical treatment standards, minimizing environmental impacts, and ensuring the welfare of animals requested that a committee formed by the National Research Council (NRC) of the National Academies explore concerns related to animal biotechnology. A committee of 12 scientists, physicians, and experts in regulatory issues accepted the task of defining science-based concerns associated with products of animal biotechnology. The committee's report presents science-based concerns it identified but it does not contain specific recommendations. Identification of the concerns will allow others to develop regulatory policy where appropriate. While the focus of the committee was on the scientific information that could clarify the issues, it remained aware of the social and other policy issues involved in moving biotechnologic advances from the laboratory to the "real world." Thus, assuming a bit of flexibility in our charge, our report addresses some of the policy issues involved as well.

In a sense, almost any issue related to a technologic advance can be a concern. The committee attempted to place concerns in relative priority order within sections of the report (i.e., hazards associated with the techniques themselves, food safety, environmental impacts, and animal welfare). In only a few cases was it possible to state that an issue brought to our table was not of concern. Much of the basic biology underlying the techniques remains to be discovered, and we have only partial information on the consequences of using biotechnologic techniques. This is true especially in terms of the environmental concerns raised. It became quickly apparent that more information was needed to assess the priority of concerns raised. Only more research will resolve this problem.

The committee relied heavily on published information, on presentations made by experts at an NRC-sponsored workshop, and on previous NRC reports. The NRC report, *Environmental Effects of Transgenic Plants: The Scope and Adequacy of Regulation*—recently completed by experts from the botanical half of the biologic world—was a valuable source of information.

This is an especially opportune time to explore the concerns related to animal biotechnology. The field has progressed to the point where we already have seen applications of this science to our daily lives, and might see many more. The committee hopes that our discussions, as reflected in this report, will inform government agencies and the public of the major scientific issues involved so that this technology can be applied as safely as possible without denying the public its benefits.

This study and the resulting report would not have been possible without the dedication, skill, and hard work of the study director, Dr. Kim Waddell, and research assistant, Michael Kisielewski, of the Board on Agriculture and Natural Resources of the National Academies.

JOHN G. VANDENBERGH, *Chair*
Committee on Defining Science-based Concerns
Associated with Products of Animal Biotechnology

Acknowledgments

This study was enhanced by the contributions of many individuals who graciously offered their time, expertise, and knowledge. The committee thanks all who attended and/or participated in its public workshop:

MICHAEL D. BISHOP, Infigen, Inc., DeForest, WI
KEITH H. S. CAMPBELL, University of Nottingham, Loughborough, UK
JOSÉ B. CIBELLI, Advanced Cell Technology, Worcester, MA
JEAN FRUCI, The Pew Initiative on Food and Biotechnology, Washington, DC
PERRY B. HACKETT, Discovery Genomics, Inc., Minneapolis, MN
MICHAEL K. HANSEN, Consumer Policy Institute, Yonkers, NY
MARJORIE A. HOY, University of Florida, Gainesville
SAMUEL B. LEHRER, Tulane University Medical Center, New Orleans, LA
LARISA RUDENKO, Integrative Biostrategies, LLC, Washington, DC
PAUL B. THOMPSON, Purdue University, West Lafayette, IN
ROBERT J. WALL, United States Department of Agriculture Research Center, Beltsville, MD

The committee extends its appreciation to the staff members of the National Research Council's (NRC) Division on Earth and Life Studies, Board on Agriculture and Natural Resources, and Board on Life Sciences for their commitment to the study process and their efforts in preparing this report.

This report has been reviewed in draft form by individuals chosen for their diverse perspectives and technical expertise, in accordance with procedures approved by the NRC's Report Review Committee. The purpose of this independent review is to provide candid and critical comments that will assist the institution in making its published report as sound as possible and to ensure that the report meets institutional standards for objectivity, evidence, and responsiveness to the study charge. The review comments and draft manuscript remain confidential to protect the integrity of the deliberative process. We wish to thank the following individuals for their review of this report:

ROY CURTISS III, Washington University, St. Louis, MO
REBECCA GOLDBURG, Environmental Defense, New York, NY
THOMAS J. HOBAN IV, North Carolina State University, Raleigh
ANNE R. KAPUSCINSKI, University of Minnesota, St. Paul
SANFORD A. MILLER, Center for Food and Nutrition Policy, Washington, DC
JAMES D. MURRAY, University of California, Davis
LARISA RUDENKO, Integrative Biostrategies, LLC, Washington, DC
PAUL B. THOMPSON, Purdue University, West Lafayette, IN
MARK E. WESTHUSIN, Texas A&M University, College Station

Although the reviewers listed above have provided many constructive comments and suggestions, they were not asked to endorse the conclusions or recommendations nor did they see the final draft of the report before its release. The review of this report was overseen by Donald D. Brown, Carnegie Institution of Washington, Baltimore, MD, and George E. Seidel, Jr., Colorado State University, Fort Collins. Appointed by the National Research Council, they were responsible for making certain that an independent examination of this report was carried out in accordance with institutional procedures and that all review comments were carefully considered. Responsibility for the final content of this report rests entirely with the authoring committee and the institution.

Contents

CONTENTS *xvii*

Tables and Boxes

TABLES

BOXES

Executive Summary

CONTEXT AND BACKGROUND

Research on genetic engineering has led to the development of a substantial variety of food and agricultural products as well as pharmaceutical and human health related products derived from several types of animals, including cows, sheep, goats, swine, fish, and insects. The federal regulatory system for genetically engineered animals and their products has been subject to increasing attention and discussion among research scientists and policymakers, as well as the public. In 2001, the Food and Drug Administration's Center for Veterinary Medicine (CVM) recognized that it was an opportune time for external scientific discussion to identify the science-based risks and concerns associated with animal biotechnology prior to any regulatory review of the food and environmental safety of these products. CVM approached the National Research Council (NRC) and requested that the NRC's Committee on Agricultural Biotechnology, Health, and the Environment convene an *ad hoc* committee of experts to identify these risks and to indicate their relative importance and potential impact.

Issues related to plant biotechnology have been addressed extensively in previous NAS reports (NRC 1989, 2000, 2002a), but a focus on animals was deemed necessary because animals have a number of unique attributes. The products of animal biotechnology, such as organs, tissues, and pharmaceuticals, can be used for direct human health needs and food. Animals present unique challenges in that they are mobile as adults and often need special care.

1

Furthermore, there is greater concern for the welfare of animals than of plants, in part because animals are considered sentient organisms.

The Task of the Committee

The specific task set before the committee was as follows:

The committee will prepare a brief consensus report to identify risk issues concerning products of animal biotechnology. Goals of the report are to (1) develop a consensus listing of risk issues in the food safety, animal safety, and environmental safety areas for various animal biotechnology product categories. These categories include, but are not limited to, gene therapy, germline modifications, knockout technologies, and cloning, (2) provide criteria for selection of those risk issues considered most important that need to be addressed or managed for the various product categories, and (3) identify and justify risk issues that were considered but not identified as important for certain product categories.

The Scope of the Report

Although future policy decisions regarding the use of animal biotechnology will no doubt take into consideration its potential benefits as well as its potential risks, the committee was not asked to examine the potential benefits. The primary charge to which the committee responded was "...to identify risk issues concerning products of animal biotechnology..." (from NRC charge above). Not all risks identified have the same importance. Because it was difficult to set overall priorities comparing risks among these areas, the committee attempted to prioritize concerns within each main area examined: food safety, the environment, and the welfare of the animals.

In its early deliberations the committee wrestled with the use of the terms "risk" and "concern". Throughout the report the committee attempts to consistently use both terms. Descriptions of "risk" are often stated and used in terms of the likelihood of harm or loss from a hazard. "Likelihood", in turn, suggests a probability, which requires that the event already has been shown to occur. The committee notes that a number of the biotechnology techniques, their applications, or products discussed in this report are still under development in research laboratories, and have not entered the environment or the food system. The term "concern", used in the title and throughout the report, is defined in the dictionary as "an uneasy state of blended interest, uncertainty, and apprehension". This definition more accurately characterizes many of the questions and issues surrounding animal biotechnology and its products that engaged the committee and resulted in this report.

Criteria for Selecting Concerns

The primary criterion for selection of concerns that emerged from committee discussions in each of these areas is based on the judgment of the immediacy and potential severity of the risk based on scientific information. The committee also categorized risks by examining a) differences between products of conventional breeding and those produced by biotechnology that might affect food safety; b) adverse effects of biotechnology on the environment in comparison to conventional techniques; c) adverse effects of biotechnology on the health and welfare of animals in comparison to conventional techniques; d) unintended genetic effects resulting from biotechnology techniques; and e) the existence of regulatory authority and the technology platform for detection and regulation of potentially hazardous biotechnology procedures.

As in any analysis of risks resulting from new technologies, it almost is impossible to state that there is "no concern" associated with an aspect of that technology. The issues identified in this report were listed as science-based concerns because the committee identified one or more outcomes that reasonably can be expected to carry some risk—even if small. Some concerns were discussed for which the committee could find no scientific basis. These were identified in the text. While the sponsor of this study is a U.S. regulatory agency, all of the concerns discussed in this report are not restricted to the U.S. and are relevant wherever this technology might be considered or applied. Finally, the committee notes that this report is "a snapshot in time"; many of the concerns and risks that are discussed are typical of any new technology, and the initial methodologies that are developed are rapidly replaced with less risky and more sophisticated techniques. It is likely that a similar rate of evolution will occur with the applications of animal biotechnology as evidenced by advances in plant biotechnology. Nonetheless, the committee often was challenged by the paucity of data that might have provided stronger insights of the relative risks for the techniques and applications that were discussed; the committee notes this point where relevant throughout the report.

INTRODUCTION

Rapid advances in biology made since the structure of DNA was clarified provide techniques that have enhanced food production and improved human health. Advances are expected to continue and are likely to have an even greater impact in the future. However, the benefits of advanced technology rarely come without attendant hazards. The focus of this report is to identify the science-based concerns related to modern, genetically-based animal biotechnology.

Biotechnology is that set of techniques by which living creatures are altered for the benefit of humans and other animals. Animal biotechnology has a long history, beginning as far back as 8,000 years ago with the domestication and artificial selection of animals. Rapid changes in animal production had been made in previous decades through procedures such as artificial selection, vaccination to enhance health, and artificial insemination to enhance reproduction. However, modern, genetically-based, biotechnology only began in the 1960s, following the discovery of the genetic code. In this report the committee moves beyond the scientific advances to examine new genetically-based technologies.

New procedures involving direct gene insertion and manipulation allow for much more rapid selection of desirable traits than older procedures. These new procedures will be described (Chapter 2) and discussed with reference to possible concerns related to the production of medical products (Chapter 3), food safety (Chapter 4), environmental issues (Chapter 5), and animal welfare (Chapter 6). The committee recognizes that the practice of biotechnology does not occur in the absence of the social, policy, and regulatory environments. Therefore, the committee concludes its report by briefly addressing these topics (Chapter 7).

During the committee's deliberations, five overarching concerns emerged. The first was whether anything theoretically could go wrong with any of the technologies. For example, is it theoretically possible that a DNA sequence from a vector used for gene transfer could escape and unintentionally become integrated into the DNA of another organism and thereby create a hazard? The second was whether the food and other products of animal biotechnology, whether genetically engineered, or from clones, are substantially different from those derived by more traditional, extant technologies. A third major concern was whether the technologies result in novel environmental hazards. The fourth concern was whether the technologies raise animal health and welfare issues. Finally, there was concern as to whether ethical and policy aspects of this emerging technology have been adequately addressed. Are the statutory tools of the various government departments and agencies involved sufficiently defined? Are the technologic expertise and capacity within agencies sufficient to cope with the new technologies should they be deemed to pose a hazard?

Among the topics considered by the committee, the effects on the environment were considered to have the greatest potential for long-term impact. The taxonomic groups that present the greatest environmental concerns are aquatic organisms and insects, because their mobility poses serious containment problems, and because unlike domestic farm birds and mammals, they easily can become feral and compete with indigenous populations.

Applications of Biotechnology Techniques

The art and science of producing genetically engineered animals have advanced very rapidly in the past few years. Production of genetically engineered animals for research purposes and commercial applications has been ongoing for approximately 20 years and is increasing in frequency and scale. Much of the early work on mammalian biotechnology is based on studies of the laboratory mouse and a few other common laboratory animals. Genetically engineered mice have become models of choice in many biomedical applications. Where appropriate, studies on laboratory animals such as the mouse are presented but are not the focus of this report. The focus is on concerns related to animal products used in agriculture and medicine.

It now is possible to generate animals with useful novel properties for dairy, meat, or fiber production, for environmental control of waste production, and for production of useful products for biomedical purposes or other human consumption. Animals also can be produced that are nearly identical copies of animals chosen for useful traits, such as milk or meat production, high fertility, and the like. A number of methods presently employed can modify the germline of various animal species for these purposes. These technologies include: introduction of new genes by transfection, retrovirus vectors, or transposons; removal or modification of genes by direct germline manipulation; and propagation by nuclear transfer of nearly identical copies of an animal. A brief description of these technologies is provided in Chapter 2, including an indication of how aspects of the procedures might result in risks. The specific concerns for risks associated with these technologies are described in subsequent chapters in which the application of the technology is described.

Several methods presently are employed for genetic engineering of various animal species. Most of these were developed originally in mouse and *Drosophila* models, and more recently have been extended to other domesticated animals. Modification of the germline of mammals can be achieved through: (1) direct manipulation of the fertilized egg, followed by its implantation into the uterus, (2) manipulation of the sperm used to generate the zygote, (3) manipulation of early embryonic tissue in place, (4) the use of embryonic stem (ES) cell lines which, after manipulation and selection *ex vivo*, then can be introduced into early embryos, some of whose germline will develop from the ES cells; and (5) manipulation of cultured somatic cells, whose nuclei then can be transferred into enucleated oocytes and thereby provide the genetic information required to produce a whole animal. The last two methods have the advantage of allowing cells containing the modification of interest to be selected prior to undertaking the expensive and lengthy process of generating animals, and greatly decreasing the number of animals used.

The technology for modifying the germline of domestic animals is advancing at a very rapid pace. Indeed, some major advances were reported during the brief period in which this report was prepared. Although many of the details of the techniques described will no doubt soon become outdated and replaced by new ones not yet considered, some general issues will remain. In particular, there will (probably) always be concerns regarding the use of unnecessary genes in constructs used for generation of engineered animals, the use of vectors with the potential to be mobilized or to otherwise contribute sequences to other organisms, and the effects of the technology on the welfare of the engineered animals themselves.

Engineering of Animals for Human Health Purposes

Genetic engineering has the potential to produce domestic animals that can be used for biomedical purposes (see Chapter 3). Such uses can be divided into three major categories: live cells, tissues, and organs for xenotransplantation; biopharmaceuticals for animal or human use; and raw materials for processing into other useful end products. The committee identified several areas of concern associated with these uses.

The development of xenotransplantation as a part of clinical practice promises great benefits in terms of making it possible to dramatically increase supplies of replacement tissues and organs where severe shortages exist today. Recipients of xenotransplanted cells, tissues, or organs, however, will be exposed to considerable risk, including the risk of novel infectious disease. Such risk is not qualitatively different from the development of other new medical procedures and might be acceptable to the recipient because of the benefits of receiving a transplanted organ. The principal concern is that the uniquely close relationship created between xenotransplanted tissue and the host will allow novel opportunities for transmission of infectious disease (e.g., one derived from porcine endogenous retroviruses, or PERVs), and possibly creation of new disease agents in the process. PERVs are of special concern since the transplant might provide the opportunity for the virus to evolve into a pathogen with the potential for transmission to others.

There is a theoretical potential for microorganisms to acquire—by recombination or transduction—genes from the vector constructs used in gene transfer. However, there is yet no uncontested example of acquisition of any gene, including drug resistance markers, by bacterial flora living in a transgenic animal. Of greater concern is the theoretical possibility of the generation of potentially pathogenic viruses by recombination between sequences of a viral vector containing a transgene and related, but nonpathogenic, viruses present in the same animal, since analogous events have been observed in the laboratory.

Although animals engineered to produce useful products will not be intended for consumption by humans or other animals, there are grounds for concern that adequate controls be in place to ensure restriction on the use of carcasses from such animals. Entry of surplus animals into the food chain poses a concern because of the possibility of people in the general population being exposed to the transgene and its expressed products.

Food Safety Concerns

The committee attempted to identify potential human health and food safety concerns for meat or animal products derived from animal biotechnology (see Chapter 4). The species considered include ruminants, such as beef and dairy cattle, sheep, and goats; poultry and eggs; swine; rabbits; and a wide array of finfishes and shellfishes. Specifically, the committee considered non-genetically engineered animals that are propagated by nuclear transfer or other cloning techniques, genetically engineered animals developed primarily for meat or animal products such as milk and eggs, and genetically engineered animals developed to produce pharmaceuticals and other medical or non-medical products. The nature of concern for all foods or food products is that they should be free of agents—chemical or biologic—which affect the safety of the food for the human or animal consumer. The committee notes that the primary food safety concern in the U.S. currently is microbial pathogens primarily originating from animal fecal contamination.

The principles for assessing the safety of food from genetically engineered animals are qualitatively the same as for non-engineered animals, but animals genetically engineered for non-food products (e.g., pharmaceuticals) might present additional concerns relating to the nature of the products which they express. Female animals might be genetically engineered to produce non-food products in their milk or eggs. The males produced through this process or the unused females might enter the food supply. The safety of food products that are derived from animals engineered for non-food purposes might present a concern. Since expression of the transgene only has limited predictability, there is a concern that the product of the transgene might enter the animal's general circulation.

A small percentage of proteins present in food can exert effects beyond nutrition, including allergenicity, bioactivity, and toxicity. The genetic engineering of animals intended for use as food will involve the expression of new proteins in animals—hence the safety, including the potential allergenicity of the newly introduced proteins, might be a concern. Allergenicity only can be reasonably assessed when the protein is known to trigger an immune response in sensitive subjects. The committee notes that the range of immune responses (allergic reactions) triggered by these novel proteins are likely to be consistent with those triggered from known allergens. The possibility that particular novel gene products might trigger allergenicity or hypersensitivity responses in some consumers will vary with the gene product at issue, and because of the potentially highly significant impacts in these individuals, poses a moderate level of food safety concern. A lower level of food safety concern exists for transgenically-derived bioactive molecules used to enhance a trait such as growth or disease resistance that could retain their bioactivity after consumption. The likelihood that a bioactive product poses harm will depend on the gene product, the food product, and the consumers involved.   However, the committee concluded that this poses a low to moderate level of food safety concern. Products that might induce toxicity are of least concern because they likely would be identified by current food safety assessment procedures.

Expression of transgenes also might result in changed nutritional attributes or improvements in the safety of food products. For example, products might include eggs that are lower in cholesterol, or meat with enhanced vitamin content or with fat content modified in quality or quantity. If these changed products were labeled in order to appeal to targeted consumers and identifiable to those who might have medical or other reasons to avoid such foods, they would be of low concern.

The committee also considered potential risks associated with cloning technologies. The cloning technologies of embryo splitting and nuclear transfer using embryonic cells were introduced into dairy cattle in the 1980s, and although they have not become widely used, over 1,400 cows were registered by the Holstein association. These cloned animals were produced to obtain more offspring from genetically valuable cows and they successfully produced calves and were milked commercially. Although there are as yet no substantial analytical studies of meat and milk composition that compare the products of the donor and the cloned animals, the milk and meat of such clones have entered the food supply, and few concerns have been raised about using these types of cloned animals for food. Based on current scientific understanding, products of embryo splitting (EMS) and blastomere nuclear transfer (BNT) clones were regarded as posing a low level of food safety concern.   Nevertheless, the committee believes that an evaluation of the composition of food products derived from cloned animals would be prudent to minimize any remaining food safety concerns. The products of offspring of cloned animals were regarded as posing no food safety concern because they are the result of natural matings.

While it is not likely that there are changes in gene expression directly related to embryo splitting or nuclear transfer that would raise nutritional or food safety concerns, the cloning of animals from somatic cells is a more recent and rapidly changing technology. This makes it difficult to draw conclusions regarding the safety of milk, meat, or other products from individuals that are themselves somatic cell cloned individuals. The key scientific issue is whether and to what degree the genomic reprogramming that occurs when a differentiated nucleus is placed into an enucleated egg and forced to drive development results in gene expression that raises food safety concerns. There currently are no data to indicate whether abnormalities in patterns of gene expression persist in adult clones and are associated with food safety risks; nor are there substantial analytical data comparing the composition of meat and milk products of somatic cell clones, their offspring, and conventionally bred individuals. Somatic cell cloned cattle reportedly are physiologically, immunologically, and behaviorally normal, and exhibit puberty at the expected age, with high rates of conception upon artificial insemination. The committee felt that it is difficult to identify concerns without additional supporting data using available analytic tests regarding food product composition. In summary, there is no current evidence that food products derived from adult somatic cell clones or their progeny present a food safety concern.

Environmental Concerns

The committee considered environmental issues to be the greatest science-based concerns associated with animal biotechnology (see Chapter 5), in large part due to the uncertainty inherent in identifying environmental problems early on and the difficulty of remediation once a problem has been identified. Any analyses of GE organisms and their potential impact on the environment needs to distinguish between organisms engineered for deliberate release and those that are engineered with the intention for confinement, but escape or inadvertently are released. The discussion in this report focuses primarily on the latter category, but the committee has a high level of concern regarding the intentional release of GE organisms into the environment. The concerns that follow primarily focus on risks resulting from GE animals entering natural environments. The release or escape of GE animals could result in a transgene spreading through reproduction with wild type individuals of the same species. The risk of horizontal gene transfer (i.e., the nonsexual transfer of genetic information between genomes by the vector) is of considerably lower probability but of high risk should it occur in some ecosystems.

The likelihood of a transgenic animal becoming established in the environment is dependent on two factors: a) its ability to escape and disperse in diverse communities, and b) its fitness in that environment. Once a transgene is introduced into a population, natural selection for fitness will determine the ultimate fate of the transgene if the population is large enough to withstand the initial perturbations. Fitness in this context refers not only to the GE organism's survival, but also to its reproductive ability, including juvenile and adult viability, age at sexual maturity, female fecundity, male fertility, and mating success (i.e., to all aspects of the organism's phenotype that affect spread of the transgene). The GE organism eventually might replace its relative or become established in that community if it is more fit than its wild relatives in that environment. If it is less fit, the engineered trait eventually will be removed from the receiving population. If the fitness of transgenic and nontransgenic individuals is similar, the likely outcome is persistence of both transgenic and nontransgenic genotypes. Transgenic organisms can be produced with changes in physiologic traits far beyond what is possible with naturally occurring mutations. For example, natural dwarfism or gigantism in mammals and poultry has effects which are limited approximately to four times the size of that of normal animals, while mean alteration in size-at-age of four to eleven times has been reported in GE salmonids. Such introduced GE animals might upset the predator–prey balance in an otherwise stable environment.

The ability of certain GE organisms to escape, disperse, and to become feral in diverse communities is of high concern. Animals that become feral easily, are highly mobile, and have a history of causing extensive community damage are of greatest concern. They include insects, shellfish, fish, and mice and rats. Mice and rats—while known to become feral easily—are not likely to escape since these transgenically altered rodents are maintained under close confinement in laboratory colonies. Animals that become feral easily, have moderate mobility, and have caused extensive damage to ecological communities raise the next most serious concerns; these include cats, pigs, and goats. Animals that are less mobile, but have been known to become feral with minimal community impact, pose the next level of concern; these include dogs, horses, and rabbits. Less mobile and highly domesticated animals that do not become feral easily, such as domestic chickens, cattle, and sheep, present the least concern, along with transgenic animals produced for human medical benefits such as xenotransplantation, which have little chance of becoming established in the environment.

Colonization by GE animals might result in local displacement of a conspecific population, which could have a disruptive effect on other species in a community. For example, the survival of predatory species that depend on a prey species eliminated by a GE organism that had become feral could be threatened. The impacts of transitory and long-term environmental harms are dependent on the stability and resilience of the receiving community. A community is deemed stable if ecologic structure and function return to the initial equilibrium following perturbation from it. These definitions allow a prioritization of potential harms from GE animals based in part on the receiving community's stability and resilience. Those that are most stable will sustain the least harm with the greatest harm occurring to unstable communities. It might be impossible to limit which communities a GE organism will gain access to; thus, if any of these communities are fragile, the concern that the GE organism will cause environmental harm will be high.

Prioritizing environmental concerns always will be on a case-by-case basis because of the uniqueness of each GE construct, founder, and receiving ecosystem. However, based on the principles of risk, the committee attempted to prioritize those concerns. Three variables were considered: (1) effect of the transgene on fitness of the animal in the environment after the escape or release of a GE animal, (2) the species transformed, and (3) stability and resiliency of receiving community. Inserting a transgene that increases fitness of a highly mobile species that becomes feral easily raises the greatest level of concern (e.g., a gene that increases salt tolerance in catfish). A transgene that increases fitness of a moderately mobile species that can become feral (e.g., the phytase gene in the pig) raises the next level of concern. Inserting a transgene that does not increase fitness in a low mobility species, which does not become feral easily (e.g. a gene for a protein of industrial value in cows), raises the least concern.

One case of immediate concern is the release of transgenic fish and shellfish. Production of some GE fish and shellfish might result in environmental benefits when compared to conventional aquacultural practices. For example, production of fish expressing a phytase transgene might allow use of less fishmeal in feeds while decreasing phosphorus in waste products from aquaculture operations. However, transgenic fish and shellfish might pose environmental hazards. Cultivated salmon have escaped into the wild from fish farms and these salmon already pose ecologic and genetic risks to native salmon stocks. In studies of transgenic salmon under laboratory conditions, some of these transgenic lines grew four to six times faster than nontransgenic salmon, with a 20 percent increase in feed conversion efficiency. In order to support their rapid growth, GH transgenic salmon consumed food at a more rapid rate than control salmon. In addition, their oxygen uptake is about 60 percent more than that of controls during routine activity and during sustained swimming. These findings suggest that the GE Atlantic salmon might show increased fitness, but gaps still exist in our understanding of the key net fitness parameters to allow an assessment of the impact of their entry into wild populations.

Possible environmental hazard pathways posed by escape or stocking of transgenic shellfish into natural ecosystems have not yet been thoroughly considered. Information is not yet available to assess ecologic risk posed by production of these organisms, but it is clear that confinement of these aquatic organisms will be difficult and they are likely to escape.

Animal Health and Welfare Concerns

The effects of genetic manipulation on animal health and welfare are of significant public concern. Animal welfare has proven difficult to assess because it is so multifaceted and involves professional and ethical judgments. The committee considered the following facets of animal welfare in discussing transgenic and cloning technologies: their potential to cause pain, distress (both physical and psychologic), behavioral abnormality, physiologic abnormality, and/or health problems; and, conversely, their potential to alleviate or to reduce these problems. Both the effects of the technologies themselves and their likely ramifications are addressed.

The applications of biotechnology can have adverse effects on the welfare of animals. For example, ruminants produced by *in vitro* culture or nuclear cell transfer methods—whether or not they carry a transgene—tend to have higher birth weights and longer gestation lengths than calves or lambs produced by artificial insemination. Large offspring syndrome (LOS) is much more frequent in cattle produced by *in vitro* techniques. Because of LOS, difficult calvings can be a problem and might require special husbandry or veterinary procedures such as caesarian sections. Additional health and welfare problems requiring special attention include respiratory distress, lack of suckling reflex, and a variety of pathologic conditions.

Some of the techniques in use are extremely inefficient in the production of transgenic animals. Efficiencies of production range from 0 to 4 percent in pigs, cattle, sheep, and goats, with about 80 to 90 percent of the mortality occurring during early development. Of the transgenic animals that survive, many do not express the inserted gene properly, often resulting in anatomical, physiologic, or behavioral abnormalities. The variability and subtly of response makes assessment difficult.

Unexpected phenotypic effects—especially on behavioral traits of genetically altered animals—might occur. Work with knockout and cloned mice has demonstrated, in some instances, elevated levels of aggression and impairment of learning and motor tasks, suggesting additional studies of cloned livestock are warranted. Although there generally are fewer potential animal welfare concerns associated with the production of transgenic farm animals for biomedical purposes than for agricultural purposes, some concerns remain. A common method to produce pharmaceuticals in animal tissues or fluids is to produce transgenic cattle or goats that express the protein of interest in mammary tissue. The recombinant protein then is secreted in milk when the female lactates. Those proteins either might be expressed in non-mammary tissues, or might "leak" out of the mammary gland into the circulation. If the protein is biologically active in the species in which it is produced, it could cause pathologies and other severe systemic effects.

An important animal welfare concern related to xenotransplantation is the management and housing of pigs intended for use as organ source animals. The pigs are maintained in sterile, often isolated environments to minimize transmission of disease to human recipients, but this environment might lead to abnormal behavioral development.

Policy and Institutional Concerns

While policy issues might be considered beyond the scope of this study, the committee took account of their existence in identifying science-based concerns about animal biotechnology. The policy framework ultimately determines the scientific questions that the regulatory process must address, and the manner in which it must address them. Although the committee's charge is limited to addressing science-based concerns, the committee notes that (1) socially, politically, and ethically determined factors influence both the nature of scientific research and the interpretation of data, (2) how one addresses scientific uncertainty or the importance of various concerns that result from introduction of a proposed technology is influenced by political and ethical considerations, and (3) technologies often have impacts on social, political, economic, religious, and spiritual conditions or values which, in turn, might impact health and the environment.

New technologies, such as biotechnology, often are characterized by a variety of uncertainties resulting in unexpected outcomes. Uncertainties can be placed in three categories—statistical, model, and fundamental. These categories of uncertainty generally correspond to technical, methodologic, and epistemologic considerations respectively, which also can be described as inexactness, unreliability, and insufficient knowledge. Regardless of the category, uncertainty also relates to the difficulty of placing potential impacts into the policy context within which proposed biotechnologies will be addressed.

Biotechnologic techniques can both impact upon, or be impacted by social, political, and ethical factors. Concern exists that certain biotechnologies can favor a particular kind of agricultural system that might induce unexpected and unwelcome changes for certain segments of the agricultural community such as small-scale farmers, or for animals or the environment. Alternatively, those changes might result in increased efficiency in food production for a growing population, improvements in animal welfare, or better protection of the environment. The socioeconomic impacts of animal biotechnologies might be manifest at the level of the individual, family, community, or corporation. For example, religious or cultural groups might have dietary norms or rules that might be violated by genetic engineering of animals used for food.

Regulatory decisions and enforcement are difficult in the absence of an ethical framework underlying regulatory decisions related to animal biotechnologies or a regulatory framework for addressing unique problems and characteristics associated with animal biotechnologies. Ethical considerations range broadly, generally are normative, and cannot be resolved scientifically. Some people, irrespective of the application of the technology, consider genetic engineering of animals fundamentally unethical. Others, however, hold that the ethical significance of animal biotechnologies must derive from the risks and benefits to people, the animals, and/or the environment. Yet another view focuses on the right of humans to know what they are eating or how their food or pharmaceuticals are being produced, and therefore labeling becomes an issue to be addressed.

The current regulatory framework might not be adequate to address unique problems and characteristics associated with animal biotechnology. The responsibilities of federal agencies for regulating animal biotechnology are unclear. How each agency will deal with scientific uncertainty remains to be seen. The committee notes a particular concern about the lack of any established regulatory framework for the oversight of scientific research and the commercial application of biotechnology to arthropods. In addition to the potential lack of clarity about regulatory responsibilities and data collection requirements, the committee also notes a concern about the legal and technical capacity of the agencies to address potential hazards, particularly in the environmental area.

The committee considers it appropriate to identify some of the potential social implications of animal biotechnology. The committee is concerned that the regulatory agencies are not clear with regard to the scope and limitation of their mandates to address such matters that do not directly affect health and the environment. Specifically, there is a need for clarity about whether the regulatory agencies consider it within their charge to consider only the direct health and environmental impacts of biotechnology, or also the social or economic impacts of a technology that, in turn, might have an adverse health or environmental impact.

1

Introduction

OVERARCHING CONCERNS OF THE COMMITTEE

Five overarching concerns of animal biotechnology dominated discussions before the committee. The first was whether anything theoretically could go wrong with any of the technologies. For example, is it theoretically possible that a vector used for gene transfer could escape and become integrated into the DNA of another organism and thereby create a hazard? The second was whether the food and other products of animal biotechnology, whether genetically engineered or not, or from clones, are substantially different from those derived by more traditional, extant technologies. A third major concern was whether the technologies raise novel environmental issues, and a fourth was whether they raise animal welfare issues. Finally, there was concern as to whether the statutory tools of the various government departments and agencies involved are sufficiently well-defined and whether the technologic expertise and capacity within agencies are sufficient to cope with the new technologies, should they be deemed to pose a hazard. Before these issues are considered in the individual chapters that follow, the committee felt that it was important to articulate how it defines "concern". The term "concern" is used throughout the report and is defined as "an uneasy state of blended interest, uncertainty, and apprehension." The committee also attempts to put the new technologies—which form the focus of this report—into perspective and to discuss some of what it has learned from past animal agricultural practices, and particularly from those technologies that have reached fruition in the past half century.

THE CURRENT STATE OF ANIMAL BIOTECHNOLOGY

Agricultural output of poultry and livestock in the United States exceeds $90 billion annually, of which around $11 billion consists of exports (USDA, 2001). There are currently about 9 million dairy cows, 5 million dairy heifers, and 85 million beef cattle and calves in the United States, and approximately 100 million hogs are slaughtered annually. However, trends in food consumption are changing. Even as the demand for red meat remains high, many consumers are changing their preferences from red meat to alternative protein sources. Americans consumed 82 pounds of chicken per capita in 2000 compared to 69.5 pounds of beef—a reversal of the situation a generation ago. Sales of farmed fish also have increased markedly as fish farming has become more productive and efficient. The main fish products traded domestically and internationally are shrimp (and prawns), Atlantic and coho salmon, and mollusks, but the market shares of tilapia, sea bass, and sea bream are increasing (Lem, 1999). Carp is, by far, the finfish type produced in largest quantity worldwide, with production about ten times that of salmon (FAO, 2000), but is primarily consumed domestically in Asian countries, rather than traded. Channel catfish constitutes the major species of finfish farmed in the United States (Lem, 1999). Per capita demand for high-quality meat and fish products is expected to increase both in response to rising world population and to improvements in the standard of living over the next 25 years (Pinstrup-Andersen and Pandya-Lorch, 1999). As a consequence of increased demand for meat and the deterioration and loss of agricultural land, there is pressure to utilize the potential for biotechnology to improve productivity in animal agriculture. As the techniques for producing transgenic animals become more efficient and as more is known about controlling how inserted genes are expressed, it is likely that the approaches soon can be integrated into agriculture. Indeed, the commercial production of transgenic fish, which is likely to occur worldwide, already is imminent.

Genetically engineered poultry, swine, goats, cattle, and other livestock also are beginning to be used as generators of pharmaceutical and other products, potential sources for replacement organs for humans, and models for human disease. The technology to produce foreign proteins in milk by expressing novel genes in the mammary glands of livestock already has advanced beyond the experimental stage, with some of the products currently in clinical trials (Colman, 1996; Murray and Maga, 1999). In theory, transgenic animals can provide milk that is more nutritious for the consumer, or that is enhanced for certain protein components that might be valuable for manufacturing cheese or other dairy products. However, the largest investments in the technology to date have been made by pharmaceutical companies interested in producing enzymes, clotting factors, and other bioactive proteins in milk.

Companies also are interested in farm animals as possible sources of replacement organs for humans. Transplantation is an accepted and successful treatment for organ failure, but there is an enormous shortage of available human organs. As there are ethical and practical concerns related to the use of donor organs from primates, the pig, in particular, is being considered as an alternative. Unfortunately, humans express antibodies to a carbohydrate epitope (terminal α1,3-galactose residues) that is present on the surface of pig cells (Sandrin et al., 1993). As a result, the xenograft immediately becomes a target for acute rejection. To remedy this situation, pigs will be produced that lack the α1,3 galactosyl transferase enzyme (Tearle et al., 1996; Dai et al., 2002; Lai et al., 2002).

Although the mouse, because of its small size, short generation times, fecundity, and well-studied genetics has become the animal of choice for providing models for human disease, farm species might provide alternatives where the mouse is inappropriate. One possible future scenario is the creation of specific gene knockouts in farm animals in order to mimic human disease in a large animal model. For example, McCreath et al. (2000), have generated genetically-engineered sheep carrying a mutated collagen gene, and have suggested that such animals could serve as models for the human connective tissue disease *osteogenesis imperfecta.*

The development of such technologies and others yet to be conceived and their incorporation into agricultural and biomedical practice raises concerns about whether the end products can be consumed safely, whether there are likely to be unwanted effects on the environment, and whether animal welfare will be adversely affected. The goal of this report is to identify concerns that will aid the federal regulatory agencies in evaluating the possibility of such adverse outcomes. However, before proceeding further, it is perhaps helpful to understand what is meant by biotechnology and to appreciate how far such biotechnology already has been incorporated into current agricultural and biomedical practice. It also is clear that the concerns of the public are focused on some of the more recent technologic advances relating to gene transfer between organisms that would not normally interbreed and to assisted reproductive procedures, such as somatic nuclear cell transfer to create so-called clones (Eyestone and Campbell, 1999; Box 1.1). Many of these recent advances have not yet left the experimental stage, but it is clear that several, including transgenic finfish, which are soon likely to be commercialized, are likely to assume importance both in agriculture and medicine.

BOX 1.1
A Definition of Cloning

The verb "to clone" and the noun "clone" have a range of meanings and interpretations. The noun is derived from the Greek word klōn, meaning a twig. Its original use in English was to describe asexually produced progeny, and it has been in familiar use in horticulture for centuries. "To clone" in this context, therefore, means to make a copy of an individual. "Clone" was later adopted into the parlance of modern cellular and molecular biology to describe groups of identical cells, and replicas of DNA and other molecules. Monozygotic twins are clones, but the term has recently become popularized in the media to mean an individual, usually a fictitious human, grown from a single somatic cell of its parent. The first reports of animal cloning were in the late 1980s and were the result of the transfer to anucleated oocytes of nuclei from blastomeres (cells from early, and presumably undifferentiated, cleavage-stage embryos), a technique that is referred to as blastomere nuclear transfer or BNT, in this report. Cloning of sheep, cattle, goats, pigs, mice, and, more recently, rabbits and cats, by transplanting a nucleus from a somatic, and presumably differentiated, cell into an oocyte—from which its own genetic material has first been removed—was achieved about a decade later (Wilmut et al., 1997; reviewed by Westhusin et al., 2001), leading to the speculation that humans also could be cloned. It is important to note that somatic cell nuclear transfer (SNT) also can be used to produce embryonic stem cells, giving researchers the opportunity to obtain undifferentiated stem cells that are genetically matched to the recipient for research and therapy, which is independent of the discussion here regarding the use of SNT for reproductive cloning of animals. Neither BNT nor SNT result in an exact replica of an individual animal, although the progeny are very similar to each other and to their donor cell parent. Any genetic dissimilarity is likely due to the cytoplasmic inheritance of mitochondria from the donor egg, which possesses its own DNA, and to other cytoplasmic factors, which seem to have the potential to influence the subsequent "reprogramming" of the transferred somatic cell genome in such a way that spatial and temporal patterns of gene expression in the embryo are affected as it develops (Cummins, 2001; Jaenisch and Wilmut, 2001). For these reasons, many scientists have objected to the use of the term clone in the context of somatic cell nuclear transfer. The committee acknowledges this shade of meaning and has attempted to make the appropriate distinction when the term *clone* is used. Nevertheless, *clone* is now so widely accepted as a synonym for somatic cell nuclear transfer—not just by the public at large—but also by embryologists and other biologists, that the committee has retained it rather than attempt to replace it with a more precise, but cumbersome, phrase.

THE ORIGINS OF BIOTECHNOLOGY IN ANIMAL AGRICULTURE

Biotechnology literally is technology based on biology; it is the application of scientific and engineering principles to the processing or production of materials by biologic agents to provide goods and services. The application of biotechnology to animals has a long history, beginning in Southwest Asia after the last ice age, when humans first began to trap wild animal species and to breed them in captivity, initially for meat and fiber and later for transport and milk. Of the approximately 48,000 mammalian species, fewer than 20 have been successfully domesticated (Diamond, 1999). Other than cats and dogs, only five of these species (cattle of the *Bos* genus, whose ancient ancestor is the now extinct auroch; sheep derived from the Asiatic mouflon species; goats, which are descended from the benzoar goat of West Asia; pigs derived from captured wild boars; and horses, which originated from now extinct wild horses that roamed the steppes of Southern Russia) are found worldwide (Diamond, 1999; Box 1.2). As pointed out by Hale (1969) and Diamond (1999), the animals that have been successfully domesticated and farmed share and exhibit a unique combination of characteristics. They are relatively docile, are flexible in their dietary habits, and can grow and reach maturity quickly on a herbivorous diet, and breed readily in captivity. They also have hierarchical social structures that permit humans to establish dominance over them, and are adapted to living in large groups. They do not include species that generally have a tendency to be fearful of humans or disturbed by sudden changes in the environment. Our ancestors no doubt based their selection methods for improving their herds and flocks on how easy the animals were to farm, as well as on potential agricultural value. In turn, the animals are adapted to thrive in a domesticated environment.

BOX 1.2
Progression of Technologies Incorporated into Modern Animal Agriculture[1]

Vaccinations and other health technologies[2]
Artificial insemination[3]
Freezing of semen[4]
Sire testing and selection[5]
Use of antibiotics in feed to increase gain[6]
Embryo transfer[7]
Embryo splitting and cloning from blastomeres[8]
In vitro maturation/*in vitro* fertilization of oocytes and *in vitro* culture of resulting embryos
Use of hormones to control ovulation in farm animals and to induce spawning in fish zygotes[9]
Hormonal sex reversal and production of monosex fish stocks
Chromosome set manipulation[10]
Steroid administration to improve weight gain
Bovine somatotropin (BST) to increase milk production in dairy cows
Marker-assisted selection

---

[1]Technologies are presented in approximate sequential order of adoption; several technologies (such as artificial insemination, which was first described in 1910 but not widely adopted until the 1950s) were developed years or decades before they were commonly used.
[2]Vaccination is used widely in the livestock and poultry industries as a protection against viral and bacterial pathogens.
[3]Artificial insemination (AI)—in conjunction with the use of frozen semen from select bulls—is common in the dairy industry but relatively rare in the U.S. beef industry. The use of fresh semen for AI is becoming increasingly important in the swine and poultry industries.
[4]Bovine semen can be successfully frozen to yield high-quality, motile sperm upon thawing. The freezing of semen is problematic for swine and other livestock.
[5]Used widely in the dairy and swine industries.

[6]Used widely to increase meat production from cattle and hogs (except in certified organic herds) *and used for pathogen control for farmed fish.*
[7]Mainly cattle, some swine.
[8]Cattle only in the U.S.
[9]These combined techniques underpin human IVF procedures, but are widely used experimentally and sometimes commercially in the livestock industry.
[10]Mollusks and finfish.

The fact that the modern breeds of these species differ so markedly from their progenitor species is a reflection of how quickly directed breeding can act. The modern Holstein, which dominates the contemporary United States dairy industry, little resembles its ancestors of only a half-century ago. Milk production per cow increased almost threefold between 1945 and 1995 (Majeskie, 1996), largely as a result of breeding from select bulls. There has been an accompanying drop in the number of cows, land devoted to dairy production and in manure produced. On the downside, the cows have a tendency towards lameness, are considerably less fertile than in the 1940s, and are frequently maintained in a herd for no more than 2–3 years or 2–3 lactations (Pryce et al., 2000; Royal et al., 2000), and represent a very narrow genetic lineage (Weigel, 2001). The export of these animals and their lineages to Europe and elsewhere is assuring the globalization of both the benefits and drawbacks of the American Holstein. Analogous changes are ongoing in the swine industry, where the pressure to produce lean, fast-growing animals of uniform size is leading to the abandonment of old breeds (Notter, 1999). Paradoxically, unless the old livestock breeds are eaten, sheared or milked, they will not survive.

The dog (*Canis familiaris*), on the other hand, provides an interesting example of the range of phenotypes that can be derived by selection within a single species. Dogs are believed to have originated in several separate domestications from wolves (*Canis lupus* and *Canis rufus*) and coyotes (*Canis latrans*) before the domestication of livestock. They have undergone remarkable modifications in size and behavior over short periods of intense selection and to provide the diversity observed in modern breeds. This reflects the enormous pool of genetic variation within the species (Wayne and Ostrander, 1999), but (possibly) also the fixation of new mutations into different genetic lineages. Inbreeding of dog breeds, as of domestic livestock, has led to a major narrowing of intrabreed variability (Zajc et al., 1997).

The same kinds of selective pressures that molded the large farm animal species has led to the creation of the modern breeds of farmed fowl, which include chickens, ducks, geese, and turkeys domesticated for their meat, eggs, and feathers. As in the dairy industry, there has been a remarkable improvement in the productivity of the poultry industry over the last 60 years. Between 1940 and 1994, yearly egg production per laying hen increased from 134 to 254, mainly as a result genetic selection. The broiler industry has shown similar gains (Pisenti et al., 1999). In 1950, a commercial bird took 84 days to reach a market weight of 1.8 kilogram. By 1988, this market weight was reached by only 43 days (Pisenti et al., 1999) on about half the amount of feed (Lacy, 2000).

Scientific breeding, combined with better nutrition and veterinary care, clearly has produced breeds of animals that are remarkably productive, although sometimes strikingly different in habits and appearance from those farmed early in the twentieth century. The practice has also led to a loss of many breeds of livestock and fowl, and a decline in genetic diversity within the breeds that survive. For example, it has been estimated that there were several hundred specialty lines of chicken in North America at the beginning of the last century, whereas the number of commercial hybrid strains now available through suppliers is fewer than 10 (North and Bell, 1990).

Aquatic animals, including finfish and shellfish, now are farmed, and specific breeds that have been selected for growth and other traits are established now in the largest industrial sectors of aquaculture, such as channel catfish, rainbow trout, and Atlantic salmon. The growth and quality of such animals are also amenable to genetic engineering through modern biotechnology. Genetically engineered or highly selected aquatic species present special problems in terms of confinement, as the features that might make them attractive commercially might pose risks to the genetic base of their wild relatives with which they can interbreed (Hallerman and Kapuscinski, 1992b).

Insects also have been domesticated for farming. The two best-known examples are the honeybee and silkworm; considerable genetic gains in productivity have provided strains of these insects far removed from the ancestral species from which they were derived. Attempts to develop strains of honeybee with improved resistance to pathogens and silkworms that produce proteins other than silk are on the horizon. Insects, like fish, are especially difficult to confine so that "escapes" are almost inevitable. In addition, insects, including ones that can be engineered transgenically, are likely to continue to be used as part of biocontrol programs for pest insects and invasive plant species and, as such, might be intentionally released into the environment. There will almost certainly be attempts to replace or to infiltrate native populations with insects that have been engineered in such a manner that they are less of a pest or unable to transmit pathogens (Hoy, 2000). Private-sector companies already have begun to farm recombinant proteins (antibodies, cytokines, enzymes, and bioactive peptides) from insect larvae. Whereas the U.S. Department of Agriculture's Animal and Plant Health Inspection Service (APHIS) regulates the release of insects for pest management, it is unclear which agency is responsible for protecting against accidental release of insects from mass rearing factories. Horizontal gene transfer, disruption of ecosystems, and native species extinctions are among the potential hazards that arise from permanent releases of transgenic arthropods into the environment (Hoy, 2000).

BOX 1.3
Examples of Technologies that are Experimentally Established but not Yet in Widespread Use in Animal Agriculture

Production of sexed semen
Production of transgenic animals by direct gene transfer
Production of transgenic animals through genetic engineering of sperm
Cloning of adult animals by somatic cell nuclear transfer to produce "copies"
Cloning of animals by somatic cell nuclear transfer to achieve genetic engineering

---

The traditional kind of biotechnology emphasized at the beginning of this section relies upon natural breeding procedures to select valuable phenotypes from the variation in the existing gene pool of a species and is beyond the purview of this report, even though it has contributed so successfully to modern-day production agriculture. It is firmly entrenched in our agricultural communities, and many are generally conversant with its benefits and risks. Importantly, other forms of research-driven biotechnologies, based on improved insight into reproductive physiology and endocrinology, embryology, genetics, and animal health also have made their way into standard farming practices over the last 75 years (Box 1.2). A few of the procedures listed extend the boundaries of biotechnology to the development of organisms that have a combination of traits generally not attainable in nature through conventional breeding and are not themselves without controversy. Some of those listed are perceived by both scientists and lay people as endangering human health or as adversely affecting animal welfare or the environment. Certain of the technologies even can have unintended, long-term consequences on the economics of agriculture itself. Finally, some of the concerns raised about the technologies in Box 1.2 are quite relevant to those listed in Box 1-3. Although several of these technologies remain experimental and have not yet become a part of standard agricultural practice, others (e.g., commercialization of transgenic fish) are undergoing government review for commercial approval. It is these newer technologies on which this report is focused. For these reasons, it is worthwhile discussing Box 1.2 and some of the issues that these technologies have raised before moving on to the ones associated with Box 1.3.

CONCERNS REGARDING EXTANT TECHNOLOGIES

Animal Health

There are well-established guidelines for the application of technologies that maintain animal health, such as standard vaccination against viral and bacterial diseases. Indeed considerable efforts are being made to expand the range of such technologies in order to prevent epidemic spread of disease in flocks and herds, which are particularly at risk when farmed under intense conditions (BBC, 2001). Even the therapeutic use of antibiotics to treat animals that have bacterial infections or are in danger of becoming infected seems not in itself to be controversial, except when antibiotics of medical importance to humans are employed.

Subtherapeutic Use of Antibiotics

The U.S. Food and Drug Administration (FDA) approved antibiotics as feed additives for farm animals in 1951. Their use since has been extended to fish farming, particularly with the global spread and dramatic increase of aquaculture in tanks and pond-like structures where antibiotics are used for prevention and control of disease rather than to enhance growth (NRC, 1999). The treated animals are found to grow more quickly and utilize feed more efficiently than animals on regular feed. At least 19 million pounds of antibiotics are used annually for subtherapeutic purposes in animal agriculture, and generally are added to feed and water (NRC, 1999). Some of these compounds, used on livestock, including penicillin, tetracycline, and fluoroquinolone used on livestock, also are prescribed to treat human illnesses, and the practice has been shown in a few instances to contribute to antibiotic resistance of human pathogens (Chiu et al., 2002; Molbak et al., 1999). It now is generally accepted in the scientific and medical communities that antibiotic resistance can be exacerbated by the widespread improper use of antibiotics. What remains controversial is whether agriculture contributes sufficiently to the problems associated with resistant pathogens to justify a complete curtailment of their use as growth promoters (DANMAP, 2000; Stephenson, 2002). A recent report from the National Research Council (NRC, 1999) failed to find a definitive link between the agricultural use of antibiotics in animal feed drinking water and antibiotic resistance of human pathogens. The report states, "The use of drugs in the food production industry is not without some problems and concerns, but does not appear to constitute an immediate public health concern." Since that report was released, additional information, raising further concerns, has been released (Fey, 2000; Gorbach, 2001). Consequently, the practice remains under intense scrutiny and is opposed by some scientific and medical organizations.

Assisted Reproductive Procedures

Artificial insemination (AI), and the later, associated use of frozen semen, sire testing and sire selection are all part of a combinatorial approach to improve the genetic quality of farmed species. AI, when first introduced into agriculture, elicited an enormous outcry from farmers, the press, and religious groups. It was claimed to be against the laws of God, a repugnant practice that would lead to abnormal outcomes, and economically unsound (Herman, 1981; Foote, 1996). It gradually has become an accepted practice in agriculture, as well as in human and veterinary medicine. The ability to freeze semen and maintain a high degree of fertilizing ability after thawing extended the power of AI, since a few select bulls could be utilized to inseminate many females in different geographic areas. Such bulls could be tested, not only for fertility, but also for their ability to sire progeny that produced copious amounts of milk. By maintaining accurate records, breeding value estimations of particular bulls could be calculated. The result was the remarkable increase in milk production, noted earlier. On the other hand, the process is leading to potentially destructive inbreeding since many of the select bulls are related. Inbreeding coefficients among modern Holsteins and Jersey breeds are now about 5 percent and rising (Weigel, 2001). The outcome might be inbreeding depression and broad susceptibility to the epidemic spread of disease. There also has been a remarkable recent loss of fertility, with successful pregnancies resulting from first insemination dropping from more than 40 percent to as low as 20 percent or less in some herds as milk yields have risen (Pryce et al., 2000; Royal et al., 2000).

Embryo recovery and transfer provides the opportunity for a particularly valuable animal to parent many more offspring in her lifetime than would be otherwise possible (Seidel, 1984). The embryos also can be frozen and then either stored or transported before they are used to initiate a pregnancy. It is a relatively common technology and has been used to produce an estimated 40,000 to 50,000 thousand beef calves every year (NAAB, 1996). The approach is to induce, by using hormones, the maturation and release of more than a single egg from the ovaries (superovulation; Driancourt, 2001). Then, the animal usually is inseminated with semen from an equally select bull, and the embryos are collected and transferred individually, or in pairs, to the reproductive tract of less valuable cows, which carry the calf to term. Modern technologies also provide the possibility of freezing the embryos and determining their gender prior to transfer. The main concern with this technique, as with the AI-associated technologies discussed above, is that it can lead to narrowing of the genetic base of the breed, in this case involving both parents. A related technique is to use a needle to aspirate immature oocytes from the ovaries (in the case of livestock the oocytes often are taken from slaughtered animals at an abattoir) and to mature the oocytes for about one day in a culture containing hormones. At the stage when the oocytes reach a point midway through the second division of meiosis, they are fertilized with live sperm. In rare instances, fertilization is achieved by a single sperm or sperm head, which is injected through the tough outer zona pellucida of the oocyte, either beneath the zona or directly into the cytoplasm (intracytoplasmic injection, or ICSI). Whatever method is used for fertilization, the resulting zygotes usually are then cultured until the embryo reaches a more advanced stage of development. In humans, of course, these combined techniques form the basis of *in vitro* fertilization procedures and have resulted in hundreds of thousands of normal infants, but the techniques also have become an important means of producing embryos for experimental purposes in agricultural research (First, 1991). Importantly, *in vitro* maturation of oocytes underpins cloning and transgenic technologies (see Chapters 2 and 6), where large numbers of competent, matured oocytes are needed to provide the many eggs necessary for nuclear transfer and pronuclear injection, respectively (see Chapter 2). *In vitro* fertilization also is used commercially to preserve the genome of particularly valuable animals that have infertility problems such as blocked oviducts or that respond poorly to superovulation (Boland and Roche, 1993), a technique described below. This commercial application of IVF is a relatively uncommon, with about 4,000 calves born from its use annually (NAAB, 1996). Few concerns have been raised about this technique, which essentially is identical to that employed for *in vitro* fertilization in humans, although some animal welfare issues have been raised (Chapter 6).

In order to manage breeding programs more intensively, control over the reproductive cycles of livestock by hormonal intervention has increased. In general the technologies are relatively benign and involve injecting the animal with hormones, usually to stop progression through the existing estrous cycle and sometimes to mimic the events that lead to selection of one or more mature follicle(s) that will ovulate. Superovulation is a technique designed to mature a cohort of follicles simultaneously, with result that several eggs are ovulated simultaneously (Nebel and Jobst, 1998; Britt, 1985). A hormone treatment analogous to that used to produce a timed ovulation in the large farm animals is used to induce gonadal maturation in fish (Mittelmark and Kapuscinski, 2001). None of these techniques have raised public health concerns, since the hormones are similar or identical to those in normal reproduction and the amounts used within the physiologic range.

Splitting or bisecting embryos became an esoteric but well-established practice in the 1980s in order to provide zygotic twins (or, in modern parlance, clones; Boland and Roche, 1993; Heyman et al., 1998). The pieces of the embryo—usually "halves," which are genetically identical in terms of both their nuclear and mitochondrial genes (see Box 1.1)—are placed in an empty zona (the protective coat around early embryos) before being transferred to different recipient mothers to carry them to term. It is estimated that only a very small number of the calves (1 to 2 percent of those resulting from embryo transfer in the United States and Canada) are produced in this manner (NAAB, 1996). Nevertheless, these animals have been introduced into commercial herds, and have produced progeny; their milk and meat are consumed by the public.

Cloning by nuclear transplantation from embryonic blastomeres (blastomere nuclear transfer, or BNT; see Box 1.1) is an expensive procedure that also has its origins in the 1970s (Willadsen and Polge, 1981; Willadsen, 1989). What distinguishes it from somatic cell nuclear transfer, the technology that led to the creation of Dolly and much of the controversy over human cloning, is the stage of development at which the nuclei are transferred (Wilmut et al., 1998). In the older procedure, the cells or blastomeres used were from the so-called morula stage of cell development (although some were from the cleavage stage and others from the blastocyst stage) when the embryo still is an undifferentiated mass and its cells presumed still capable of forming all tissues of a fetus.

The cloning technologies of embryos splitting (EMS) and embryonic nuclear transfer (NT) were introduced into dairy cattle breeding in the 1980s. The Animal Improvement Programs Laboratory of the USDA's Agricultural Research Service (ARS) is responsible for tracking the performance of dairy cattle throughout the U.S. Recently, working with the Holstein Association, they evaluated the performance of cloned Holsteins produced by EMS and NT (H.D. Norman, USDA–ARS, personal communication). The numbers of EMS and NT clones were documented by gender and birth year. All NTs were from embryos rather than adult cells. Through 2001, there were a total of 2,226 EMS (754 males and 1,472 females) and 187 NT (61 males and 126 females) Holstein clones registered. Of female EMS clones, 921 had yield records, and 551 had noncloned full siblings with yield records. Of the 126 female NT clones, 74 had yield records, but only 11 had noncloned full siblings. These familial relationships were used to compare the performance of cloned and noncloned full siblings for standardized traits and genetic evaluations as part of the national evaluation program. These standardized traits included total milk yield, fat content (by weight and pecent), protein content (by weight and percent), somatic cell score, and productive life (in months). Also calculated were yield from contemporaries and predicted transmitting ability. Norman and his colleagues concluded that the numbers of clones have decreased for EMS males and for all NT clones over the past decade. Animals that were selected for cloning were slightly superior genetically to the contemporary population mean for yield traits; the yields of NT clones were similar to, and those of EMS clones were slightly less than, those of their noncloned full siblings.

"Modern" cloning involves taking an unfertilized egg, removing its chromosomes, and introducing the nucleus from a differentiated cell of the animal to be cloned, which is frequently an adult (Box 1.1; Wilmut et al., 1997; Polejaeva et al., 2000; Kuhholzer and Prather, 2000). The introduced nucleus is reprogrammed by the cytoplasm of the egg and directs the development of a new embryo, which is then transferred to a recipient mother to allow it to develop to term. The offspring formed will be identical to their siblings and to the original donor animal in terms of their nuclear DNA, but will differ in their mitochondrial genes and possibly also in the manner their nuclear genes are expressed or biochemically engineered (see Box 1.1 and Chapter 2). Cloning from blastomeres, the older of the two procedures, has been reported to result occasionally in large calves (and lambs), the so-called large offspring syndrome (LOS; Young et al., 1998; Sinclair et al., 2000). Analogous, though possibly more serious, abnormalities might be associated with cloning from somatic cells and are discussed further in Chapters 2 and 6 of this report.

Hormone-Treated Cattle

Among the most contentious technologies used in animal agriculture is the use of steroid hormones to increase the rate of weight gain and to reduce accumulation of fat deposits of young heifers and steers as part of the "finishing" process prior to slaughter (Heitzman, 1976; Lammers et al., 1999). The steroids are administered by slow release from a plastic implant embedded beneath the skin of the ear, which provides "physiologic" circulating levels of the hormone in the bloodstream. The hormones used are mainly Zeranol (in Ralgro™), a naturally occurring fungal metabolite (zearalenone) with estrogenic action; estradiol, progesterone, and testosterone, or mixtures of these steroids (in various Synovex™ formulations); and trenbolone (Doyle, 2000). Concern about these hormones is probably, in part, a legacy of diethylstilbestrol, which was eventually banned from use in the poultry and beef industry because of its adverse effects on humans. However, the amounts of present-use compounds consumed from meat derived from treated cattle are small, and numerous scientific studies generally have indicated that these residues exist at such low concentrations that they pose little risk to consumers (Doyle, 2000; Lange et al., 2001; United States Mission to the European Union, 1999; Henricks et al., 2001), provided good veterinary practices are employed (e.g., using the correct number of implants and placing implants correctly in the ear cartilage), although the U.S. Geological Survey has recently documented the presence of hormones in a number of streams and rivers (some of these hormones likely come from implants; Kolpin et al., 2002). Despite the scientific evidence for safety, the European Union implemented a ban on U.S. beef imports, valued at over $100 million per year in 1989 (Andrews, 1997). A concern that has not been extensively examined so far is whether these hormones pose any sort of environmental threat through their leaching into soil and water. For example, two recent studies have shown that a commonly used androgenic growth promotor—trenbolone—has been found in groundwater near cattle feedlots, and that this growth promotor has androgenic effects (Gray, et al., 2001; Schiffer, et al., 2001).

Bovine Somatotropin

The use of bovine somatotropin (BST) to increase milk yield from dairy cows has had a similar checkered history and is the subject of trade disputes. Currently banned in Europe even for experimental studies, BST was approved by the FDA for use in U.S. dairy cattle in 1993 because testing had revealed no concerns regarding consumer safety (Juskevich and Guyer, 1990; Bauman, 1999). The Monsanto product, Posilac™, now is widely used throughout the U.S. dairy industry, where milk production can be increased as much as 30 percent in well managed, appropriately fed herds, without adversely affecting the quality or composition of the milk. The BST, which is almost indistinguishable in sequence from the natural hormone, is present in low concentrations in milk, but has no biologic activity in humans. The level of IGF-1, the hormone induced by BST, is somewhat elevated but within the "physiologic range" for cows and is probably digested along with other milk proteins in the adult stomach, although it might have biologic activity in the intestine of neonates (Burrin, 1997). In its assessment, the FDA did not report that BST or IGF-1 pose any risk either in humans or animals that consume cows' milk. As with other technologies that increase productivity, a concern frequently raised is why more milk is needed when the developed world appears to have more than enough of the product. One answer is that increased productivity translates into fewer animals, producing less waste and utilizing less land—an extremely important consideration for future land management use.   The greatest concerns about BST are probably in the area of animal welfare. High-yield milking cows show a greater incidence of mastitis than lower-producing cows, but studies have shown that mastitis is not exacerbated by BST administration (Judge et al., 1997). Another concern—a practical one for the dairy industry—is a recent trend to breed heifers only once and then to sustain milk production for as long as 600 days by using BST. Lengthening lactation via BST in second calf and older cows is a larger contributor to having fewer calves per lifetime in the herd than first-calf heifers. The result has been a shortage of replacement heifers for producers, since only one calf is born during the milking life of the animal (Harlow, 2002).

Marker-Assisted Selection

Marker-assisted selection involves establishing the linkage between the inheritance of a particular trait—which might be desirable, as in the case of milk yield—or undesirable, as in susceptibility to a disease, with the segregation of particular genetic markers. Thus, even if the gene that controls the trait is unknown, its presence can be inferred from the presence of the marker that segregates with it. This technology, which is particularly important for studying complex traits governed by many genes, has only recently become a factor in animal breeding and selection strategies (Georges, 2001). Its use likely will increase exponentially as the industry incorporates the data from the various genome sequencing projects and as the density of useful, segregating markers increases on the chromosomes of the species. Initially, animals will be screened for genes that control simple traits, such as horns, which are undesirable in cattle, and halothane sensitivity, which segregates with metabolic stress syndrome in pigs. With time, easily identifiable markers will be chosen that accompany the many genes controlling more complex traits such as meat tenderness and taste, growth, calf size, and disease resistance. The approach has enormous potential for improving the quality of agricultural products, disease resistance, and other traits but could be misused (Dekkers and Hospital, 2002). For example, stringent selection of prime animals could potentially narrow genetic diversity even more than is evident at present. Use of the technique also could maximize short-term gain in productivity but at the expense of longer-term improvement due to what has been termed polygenic drag (Dekkers and Van Arendonk, 1998; Dekkers and Hospital, 2002). In essence, the cumulative effect of genes with effects too small to be exploited in a marker-assisted selection program could contribute more to increasing desired traits than genes with major effects. However, marker-assisted selection might be a powerful measure to counter inbreeding by providing genetic measures of heterozygosity, encouraging breeding strategies that maintain diversity at the majority of sites in the genome, and allowing the genetic potential of rare breeds and wild ancestors to be utilized and incorporated into mainstream agriculture.

Chromosome Set Manipulation in Mollusks and Finfish

Altering the chromosome complement of an animal can be a useful way of rendering that animal infertile, and is exploited widely in the production of fish and mollusks. Well-timed application of high or low temperatures, certain chemicals, or high hydrostatic pressure to newly-fertilized groups of eggs can interfere with extrusion of the second polar body (the last step in meiosis), resulting in "triploid" individuals with three, instead of the usual two, chromosome sets (e.g., for oysters; Allen et al., 1989). A later treatment can suppress the first cell division of the zygote, resulting in "tetraploid" individuals with four sets of chromosomes. Crossing tetraploids, which are fertile in some species, with normal diploids can then produce large numbers of triploids (Scarpa et al., 1994). Such chromosome set manipulations have been applied to cultured marine mollusks to produce confined stocks of triploids that are unable to reproduce. This application is of particular importance, as some of the shellfishes most suited to aquaculture are not indigenous to a given area and can pose ecologic risks to native species should they or their larvae escape confinement and enter natural ecosystems (USDA, 1995). Induction of triploidy reduces the likelihood that an introduced species would establish self-sustaining populations, because such animals are theoretically sterile. For example, the triploid Suminoe oyster (*Crassostrea ariakensis*) is being assessed for oyster production in the Chesapeake Bay, where diseases complicate restoration of the native Eastern oyster (*C. virginica*). Should triploidy prove an effective means for reproductive confinement, culture of sterile Suminoe oysters could support the recovery of the declining Chesapeake oyster production industry.

Another benefit of producing sterile mollusks is in maintaining product quality throughout the year. The meat quality of oysters is high just before they spawn, but low after spawning. The product quality of reproductively sterile, triploid oysters remains high year-round. Hence, triploid stocks of Pacific oyster (*Crassostrea gigas*) provide a tangible benefit to aquaculturists, and now make up almost half of commercial production in the Pacific Northwest.

Unfortunately, repeatable induction of 100 percent triploidy on a commercial scale poses a considerable technical challenge. Non-triploid larvae within batches of larvae easily can go undetected if their frequency is low (USDA, 1995). Should triploidy be desired for purposes of maintaining product quality and the species is indigenous to an area, no harm is posed. If, on the other hand, triploidy is to be utilized for reproductive confinement purposes, the presence of reproductively fertile individuals—even in low numbers—might establish progeny and a self-sustaining population. There also are indications that a small percentage of triploid oysters can progress to a "mosaic" state, with diploid cells arising within the background of triploid cells, leading to the possibility that they could produce viable gametes (Calvo et al., 2001; Zhou, 2002).

Triploidy often has been used to reduce the likelihood that introduced finfish species would establish self-sustaining populations. Use of all-female triploid stocks has been suggested as a means of achieving reproductive confinement of transgenic fishes, including Atlantic salmon (the leading candidate for commercialization). As with mollusks, however, repeatable induction of 100 percent triploidy poses a considerable technical challenge, and commercial net pen operations produce hundreds of thousands of salmon, with many escaping (Hallerman and Kapuscinski, 1992b; Carr et al., 1997; Fiske and Lund, 1999; Volpe et al., 2000).

Another technology used on finfish is to farm monosex fish stocks (Beardmore et al., 2001), which are preferred by producers either because one gender grows faster or larger than the other (e.g., males in catfish and tilapia, females in rainbow trout), or because certain species (e.g., tilapia) attain sexual maturity before reaching harvest size. Monosex populations have been established in several ways, but most reliably through hormone-induced gender reversal. All-male fry can be produced by direct administration of testosterone in feed, or all-females by administration of estrogens. Monosex stocks also can be produced indirectly by gender reversal and progeny testing to identify XX males for producing all-female stocks, as in trout (Bye and Lincoln, 1986) and salmon (Johnstone and Youngson, 1984), or YY males for producing all-male stocks, as in tilapia (Beardmore et al., 2001).

LIMITS OF THE REPORT

The above examples illustrate that a spectrum of earlier biotechnologies already has become integrated into agricultural practice. The introduction of new technologies does not mean that there are no concerns or even dangers posed by their use, or that there is universal acceptance among the public. The experience of the last 50 years, if nothing else, illustrates that there must be continued vigilance even after technologies have been approved. Conversely, it should be recognized plainly that increases in agricultural efficiency brought about by new technologies, such as those discussed above, undoubtedly have contributed to a more abundant, cheaper, more varied and lower cost food supply, and to enormous savings in agricultural land use.

Some technologies in Box 1.2 bridge the gap between what is an already established commercial practice and what is new (Box 1.3). For example, cloning from blastomeres (Box 1.1) in reality is little different from nuclear transfer from somatic cells, listed in Box 1.3, except that the transferred nuclei might not have to be so extensively reprogrammed in the cytoplasm of the recipient oocyte. Similarly, chromosomal set manipulation remains partly experimental and partly an active commercial technology.

Box 1.3 is a partial list of technologies that either are very close to being commercially available (pending approval from regulatory agencies) or are predicted to emerge from experimental to commercial use quite soon. The first one listed, the production of single sex sperm, is achieved through a cell sorting procedure that depends upon the higher DNA content of female sperm (Johnson, 2000; Lu et al., 1999). The technology is not expected to raise any new concerns and, provided the procedure can be scaled up, is likely to be highly beneficial in the dairy industry, where there is a surfeit of low-value bull calves, and to the beef industry, where males have a higher production value than females. The remaining technologies, however, might be more worrisome to the public and to the regulatory agencies, and it is these that are addressed in this report.

In terms of the types of technologies discussed, the scope of the report had, of necessity, to be limited. Three criteria are emphasized in this report:

1. The first criterion is immediacy of technologic commercialization, particularly if the products already are impinging on the regulatory system. It is clear that some of these technologies (e.g., commercial production of transgenic finfish) already are beyond the experimental stages of development. In addition, some biopharmed

drugs are in Stage 3 clinical trials and decisions must be made soon about the disposition of the livestock involved.

2. A second criterion is the potential impact of the technology. Some new procedures seem unlikely to raise concern (e.g., the sperm sexing discussed in the previous section) or represent relatively minor changes in practice. Other technologies might be broadly adopted, yet the possible harm they could cause and the overall benefits to society are difficult to evaluate.

3. A third criterion is whether there is sufficient information available about the technology to evaluate concerns properly. Indeed, the committee explicitly acknowledges that there are uncertainties associated with the application of each of the technologies discussed in this report. Unresolved scientific uncertainty interferes, not only with attempts to determine how best to apply emerging technologies to animals, but also how to predict the impacts of their application. Some hazards (see Box 1.4) remain theoretical, uninvestigated, poorly characterized, or even unknown. Such uncertainties present significant challenges to scientists and policy makers who wish to estimate the likelihood and distribution of harms and benefits resulting from application of those technologies. For example, some outcomes of applications of the technologies listed in Box 1.3, such as production of transgenic animals by gene transfer, are very difficult to predict. Uncertainties range from mere inexactness and unreliability to those that are fundamentally unknowable *a priori* (Funtowicz and Ravetz, 1992). Clearly, technologies that pose high stakes and high uncertainties pose fundamentally different challenges than those posing low stakes and little uncertainty. For this reason, for each concern discussed in this report, the committee has attempted, where possible, to specify (1) what is known, (2) the certainty with which it is known, (3) what is not known, (4) what is suspected, and (5) the limits of the science.

The committee also recognizes that there likely are either species or categories of species of animals not discussed specifically regarding concerns associated with biotechnology. Two examples of categories include companion animals and wildlife. While there are likely to be unique concerns that emerge with both categories, the concerns identified in the report regarding applications of the technologies (Chapter 2), environmental issues (Chapter 5), and animal welfare issues (Chapter 6) are all relevant and should be included in any considerations of wildlife and companion animal species.

BOX 1.4
Harms, Hazards, and Risks

The charge of the committee was to identify, but not to quantify, risk issues concerning products of animal biotechnology, and to provide criteria for selection of those risk issues considered most important that need to be addressed or managed for the various product categories. In order to provide criteria for selection of risk issues, it is important to understand how risk is determined. As outlined in Chapter 5 and as set forth by NRC (1983; 1996), a *hazard*: is an act or phenomenon that has the potential to produce harm, and *risk* is the likelihood of harm resulting from exposure to the hazard. This committee used the NRC (1996) definition of risk to develop a set of working steps to prioritize concerns. Because risk is the product of two probabilities: the probability of exposure, and the conditional probability of harm given exposure has occurred, the steps in risk analysis are to: (1) identify the potential harms, (2) identify the potential hazards that might produce those harms, (3) define what exposure means and the likelihood of exposure and 4) quantify the likelihood of harm given that exposure has occurred. (The committee notes that risk analysis in other fields can and does include additional steps in risk assessment; see Kapuscinski, 2002). Multiplying the resulting probabilities then was used to prioritize risk. While absolute probabilities are difficult to determine at this time, relative rankings from high to low are possible based on available evidence for each category. The risks, harms, and hazards are different for each chapter because the issues are different (i.e., a hazard resulting in an animal wellbeing concern might not be an environmental or human health concern).

Discussion of concerns regarding impacts of GE mice on the environment and human health also are limited in this report for several reasons. GE mice are not part of the animal production system for human food, and laboratory mice are highly unlikely to escape the confines of animal facilities because of their economic value and the generally high-quality care given to laboratory rodents. While mice might be a high risk for escape, might feralize easily, and might carry many different transgenes, the functionality of the transgenes used in mice rarely has been for a construct that will increase fitness in natural environments. Thus, the overall risk for most constructs is expected to be low. If mice were developed to be resistant to pest control measures (pesticides) or to be more disease resistant, then risks would be much higher. However, the use of mice in this way seems quite unlikely.

2

Applications of Biotechnology Techniques

INTRODUCTION

The art and science of producing genetically engineered animals have advanced very rapidly in the past few years. It now is possible to generate animals with useful novel properties for dairy, meat, or fiber production, for environmental control of waste production, for biomedical purposes or other human consumption, or that are nearly identical copies of animals chosen for useful traits, such as milk or meat production, high fertility, and the like. This chapter addresses the current state of the art of these technologies and then point out specific concerns that arise as a consequence of their application. Subsequent chapters will discuss how the technical issues can directly affect human health, the food supply, animal welfare, and the environment.

INTRODUCTION OF NOVEL GENES

A number of methods are presently employed for genetic engineering of various animal species. Most of these were developed originally in mouse and *Drosophila* models and have only more recently been extended to other domesticated animals. Access to the germline of mammals can be obtained by: (1) direct manipulation of the fertilized egg, followed by its implantation into the uterus; (2) manipulation of the sperm used to generate the zygote; (3) manipulation of early embryonic tissue in place; (4) the use of embryonic stem

(ES) cell lines which, after manipulation and selection *ex vivo*, can then be introduced into early embryos, some of whose germline will develop from the ES cells (Smith, 2001); and (5) manipulation of cultured somatic cells, whose nuclei then can be transferred into enucleated oocytes and thereby provide the genetic information required to produce a whole animal. The last two methods have the advantage of allowing cells containing the modification of interest to be selected prior to undertaking the expensive and lengthy process of generating animals. Usable ES cells are not available for all species of interest, however, and generation of embryos by nuclear transfer (NT) from somatic cells is becoming the method of choice for genetic engineering and duplication of nearly genetically identical animals (Westhusin et al., 2001).

Manipulation of the avian germline is difficult since ES lines are not available and the early embryo is difficult to access. Much current work focuses on the use of blastodermal cells or primordial germ cells, which can be cultured briefly and manipulated to modify the germline prior to introduction into fresh embryos to create chimeras from which modified lines can eventually be developed, albeit with low efficiency (Aritomi and Fujihara, 2000).

There are two basic approaches presently in use for inserting DNA into vertebrate germline cells, transfection and infection with retrovirus vectors. A third approach, based on the use of mobile genetic elements, has been commonly used for insects and is being explored for germline modification of vertebrates (Izsvak et al., 2000).

Transfection

Transfection methods include: (1) direct microinjection of DNA into the cell nucleus; (2) electroporation—introduction of DNA through transient pores created by controlled electrical pulses; (3) use of polycations to neutralize charges on DNA and the cell surface that prevent efficient uptake of DNA; (4) lipofection, or enclosure of DNA in lipid vesicles that enter a cell by membrane fusion much in the manner of a virus, and (5) sperm-mediated transfection, possibly in conjunction with intracytoplasmic sperm injection (ICSI) or electroporation (see Chapter 6). The manner of introduction of DNA is a technical issue, determined empirically for each system, and makes little difference to the final outcome. In general (with the exception of homologous recombination, discussed below), the structure of DNA introduced into a cell by any of these methods is highly variable and uncertain. Often, only a fragment of the transfected DNA is integrated into the chromosome, frequently in multiple copies, that often are integrated in long tandem arrays (Gordon and Ruddle, 1985). When transfecting cultured somatic or ES cells, a selectable marker, such as the gene encoding phosphotransferase, is often included as part of the DNA to permit selection for its presence either in eukaryotic cell lines or in the bacteria in which the DNA was mass-produced.

Retrovirus Vectors

Retroviruses are infectious elements that replicate by a unique process involving copying of the viral RNA genome into DNA (a process called reverse transcription) followed by its specific and stable introduction into host cell DNA (integration). The integrated DNA then can be expressed using the normal transcriptional machinery of the cell. Retroviruses commonly are used to introduce genes of interest into cells in culture or into somatic tissue in experimental animals (Miller, 1997). They also have been used for germline modification of fish (Amsterdam et al., 1997), mollusks (Lu et al., 1996) chickens (Thoraval et al., 1995), mice (Soriano et al., 1986), and, more recently, cattle (Chan et al., 1998). To make a retrovirus vector, a DNA construct containing the gene of interest is flanked by sequences necessary for replication as a virus. These sequences include transcriptional promoters in the long terminal repeats (LTR's), which flank the integrated DNA, or provirus. Signals necessary for packaging of the transcript in virions (virus particles), for reverse transcription, and for integration of the resulting DNA also must be included. Introduction of such a DNA construct into cells that express viral proteins, but that are incapable of making infectious virus (i.e., helper, or packaging, cells), leads to the creation of infectious virions containing an RNA copy of the gene of interest. After infection of cells with such virions, the RNA is copied into DNA and integrated at random sites in the cell genome. Again, selectable markers often are included in the construct to select cells containing the desired virus construct.

Transposons

Transposons are DNA elements that (in the presence of the appropriate gene products, or transposases) can transfer their information from one site to another in the same cell. A variety of transposons have been found in insects (Handler, 2001) and fish (Ivics et al., 1997), and some are routinely used as vectors for the generation of transgenic insects (Braig and Yan, 2002). No active transposons of these types have been observed in mammals, although the human genome contains thousands of copies of a DNA sequence related to the *mariner* transposon of *Drosophila* (Lander et al., 2001), suggesting that there might have been active elements in our recent evolutionary history. Nevertheless, several recent reports suggest that naturally-occurring transposons found in insects, or even bacteria, might provide a useful and efficient means of introducing genes into the germline of animals. *Mariner*, for example, has been shown to be active in chick zygotes, transferring its DNA from a microinjected plasmid into the germline, albeit at low efficiency (Sherman et al., 1998). A modified version of *Sleeping Beauty*, a related element from fish, has been developed to give a high yield of germline or somatic transformants in cultured cells (Ivics et al., 1997) and laboratory mice (Personal communication, P. Hackett, University of Minnesota). In practice, it remains to be seen whether—and how efficiently—genes of interest can be transferred in this way. Another transposon system being investigated for this purpose is the T DNA of *Agrobacterium tumefaciens*, a natural pathogen of plants. This bacterium can fuse with plant cells, leading to transposition of the DNA (along with whatever genes it carries) into the nuclear DNA of the host. This technique is widely used for the generation of transgenic plants (Halford and Shewry, 2000). Remarkably, it recently has been shown that *Agrobacterium* can fuse with human cells in culture, leading to transfer of T DNA carrying a marker gene (Kunik et al., 2001). None of the transposon-based techniques currently are used for the generation of transgenic livestock, but they might lead to more efficient methods for this purpose.

DIRECTED GENETIC MANIPULATION

Another goal of transgenic technology is the creation of engineered animals that lack specific genes (knockout), or have these genes replaced by one that has been engineered in a specific way (knockin; see Box 2.1). For example, transplantation of organs or tissues from non-primates (such as pigs) to humans (xenotransplantation) is currently impossible, due in part to a dramatic ("hyperacute") immune response by human recipients to a carbohydrate on the surface of pig cells (galactose-1,3-galactose); this carbohydrate is not found in old-world primates (Galili, 2001). Inactivation of the enzyme (galactosyl transferase, GT) in donor pigs could alleviate this problem, and pigs with one allele of the gene encoding this enzyme recently have been produced (Lai et al., 2002), giving rise to expectation that completely GT-deficient animals soon will be available. Another important goal is to eliminate from cattle the gene encoding prion-related protein (PrP), the protein associated with scrapie in sheep and bovine spongiform encephalopathy (BSE, or mad cow disease). Removal of this gene from mice has, at most, subtle phenotypic consequences, yet renders them completely resistant to these diseases (Bueler et al., 1992). If the mouse model holds true in cattle, homozygous knockout of bovine PrP could lead to the elimination of BSE.

BOX 2.1
Knockout and Knockin Technology

In order to study the relationship between proteins and gene function, scientists now can prevent the manufacturing of a protein by a specific gene. By disabling a gene from a test organism, and then producing descendants that contain two copies of the disabled gene, it is possible to observe the descendants' development in the absence of a particular protein. This practice, referred to as *knockout* technology, is an attempt to shut down or turn off a particular gene. Thus far, the mouse has been the mammal in which knockout technology has been most generally applied (University of Guelph, 2001). In essence, a "knockout" organism (e.g., the mouse) is created when an embryo cell (an embryonic stem cell—or ESC—which is a cell that has yet to divide into different tissue cells; NRC, 2002b) is genetically engineered, and then inserted into a developing embryo. The embryo then is inserted surgically into the womb of a host (e.g., a female mouse). Once the embryo has matured, a portion of its stem cells will produce egg and sperm with the knocked-out gene.

A gene also can be altered in function, in contrast to being deleted. When a gene is altered but not shut down, a "targeted mutation" effect is created. This practice is referred to as *knockin* technology, whereby a life form has an altered gene "knocked" into it (MGD, 2002).

Gene knockout/knockin technology is well established as an experimental tool in mice due to the availability of ES cell lines. The principleis to take advantage of a rather rare event that occurs after introduction of DNA into cells—homologous recombination between identical sequences in the genome and the transfecting DNA (Bronson and Smithies, 1994). In the most common protocol, a selectable marker (such as the neomycin resistance gene) is inserted within a piece of DNA corresponding to a portion of a gene of interest. After transfection of cells by this construct and selection for the marker (by growth in a medium containing the neomycin-related antibiotic G418, in this example), the selected cells are screened to identify the small fraction that has one copy of the gene of interest disrupted by the marker. Progeny animals derived from the cells will be heterozygous for the "knocked-out" or "knocked-in" gene; breeding to obtain homozygous animals is straightforward. Because the process is so inefficient, very large numbers of transfected cells must be screened, making the use of cultured cells essential, since it would be impracticable to screen large numbers of progeny from microinjected eggs. The galactosyl transferase-knockout pigs discussed above were generated from cultured fetal fibroblasts manipulated in this way. Nuclei from these cells then were transferred into oocytes as described in the next section.

PROPAGATION BY NUCLEAR TRANSFER

In February of 1997, Dolly the sheep was introduced (Wilmut et al., 1997) and the public subsequently was inundated with opinions about the power and potential of creating new animals from somatic cells. Dolly represented the most recent advance in genetic technology—the production of multiple individuals nearly genetically identical to an adult animal. In this technique, somatic cells from an appropriate tissue are grown in culture and their nuclei are injected into enucleated oocytes obtained from another individual of the same or a closely related species. After a further period of culture, the partially developed embryos are transferred into a foster mother. This technology is being developed rapidly for many species of interest (Table 2.1) and promises to become a rapid and efficient means of propagating domestic animals with desired traits, whether those are naturally derived and selected or genetically engineered (Betthauser et al., 2000; Lanza et al., 2001; Westhusin et al., 2001). The process often is referred to as "cloning" (see Chapter 1). The nuclear transfer technique was based on previous studies in frogs conducted during the previous 5 decades (Briggs et al., 1951; Prather et al., 1999), but, until Dolly, it was unclear whether nuclei from highly differentiated somatic cells could be reprogrammed to a pattern of gene expression suitable for directing normal development of a mammalian embryo.

TABLE 2.1 State of the art of transgenic technology for selected organisms.

Organism	Transfection	Viral vectors	Transposon	ES cells	Nuclear transfer
Mouse	4[a]	2	1	4[a]	2
Cow	3	1	0	0	2
Sheep	3	0	0	0	2
Goat	3	0	0	0	2
Pig	3	0	0	0	2
Rabbit	3	0	0	1	0
Chicken	1	2	1	0	0
Altlantic salmon	3	0	0	0	0
Channel catfish	2	0	0	0	0
Tilapia	3	0	0	0	0
Zebrafish	1	0	0	1	1
Crustaceans	1	1	0	0	0
Mollusks	1	1	0	0	0
Drosophila	2	2	2	2	0
Mosquito	1	0	2	0	0

NOTE: 0: No significant progress.
        1: Has been accomplished experimentally (proof of concept).
        2: Routine experimental use.
        3: Commercialization sought.
        4: Widespread production.
        [a] For experimental uses.
        See (Dove, 2000)

The ability to reprogram the nucleus of donor cells for successful nuclear transfer appeared initially to require the use of methods that facilitate cell cycle synchrony (Stice et al., 1998), since only after the donor cells were induced to become quiescent could offspring be obtained by transfer of the donor nucleus to enucleated oocytes (Wilmut et al., 1998). The necessity for quiescence is not as clear today, since nuclei from actively dividing cells have now also been used successfully for this purpose (Cibelli et al., 1998; Kasinathan et al., 2001; Kuhholzer et al., 2001). The ability to reprogram the donor cells also depends on the species and nuclear transfer procedure. One hypothesis is that differences in timing of embryonic genome activation contributes to differences in cloning efficiency among species (Stice et al., 1998). Currently, only oocytes can be used successfully, as they are the only recipient cells that convert differentiated nuclei into undifferentiated stages resembling pronuclei in freshly fertilized zygotes, a step which is essential for the complete development of the reconstructed embryo (Campbell, 1999; Fulka et al., 2001). How the enucleated oocyte (cytoplast) accomplishes this reprogramming is currently unknown.

At present, propagation of animals by nuclear transfer is inefficient, with an average of less than 10 percent of the embryos resulting in live offspring, although the success rate appears to be increasing with experience (Cibelli et al., 2002). Most of the failures occur during development (most often in the first third of the pregnancy for cattle and sheep), and there appears to be an increased rate of perinatal death relative to normally-conceived offspring. In cattle, at least, the developmental and perinatal problems appear to be as much a function of the *in vitro* culture technology as of the nuclear transfer itself (see Chapter 6). However, even with this existing low efficiency, there are many potential applications for reproducing highly desired genotypes, including rare or endangered species, household pets, elite sires or dams, breeds with desirable production traits but low fertility, sterile animals such as castrates and mules, or transgenic animals that have high value and for which rapid propagation is desirable. Another important application of this technology is in the dissemination of germplasm as embryos and consequent reduction of the associated risk of disease spread (Prather et al., 1999). It also is important to note that there are significant differences between cattle and swine in terms of the utility of this technique. In cattle, the ejaculate from a single bull can be used to breed 400 to 500 females in AI programs. In contrast, the ejaculate from a single boar can be used to breed only 10 to 20 females. Thus embryos obtained from NT might be the method of choice for the dissemination of swine germplasm rather than AI (Prather et al., 1999). A number of variables influence the success rate of nuclear transfer. These include: species, source of the recipient ova, cell type of donor nuclei, treatment of donor cells prior to nuclear transfer, and the techniques employed for nuclear transfer (Westhusin et al., 2001).

TECHNICAL ISSUES WITH GERMLINE MODIFICATION

Expression of Randomly-Inserted Genes

A key consideration in development of transgenic animals is ensuring that the gene product of interest is expressed in the correct tissue and at the appropriate level and time. Specifics of how this goal is accomplished vary considerably from one system to another, but some general principles can be elucidated. First, the natural regulatory elements (promoters) for most genes that direct tissue-specific transcription are complex, large, and poorly understood. For this reason, well-characterized promoters for other genes are appended to DNA encoding the desired gene product. Second, the expression of transgenes, especially those derived by transfection, is strongly under the influence of control elements in the DNA around the integration site (Wolf et al., 2000). These positional effects often lead to silencing of the gene of interest (or other genes near the integration site), or, more rarely, to unregulated expression (Bonifer et al., 1996; Henikoff, 1998). They can be alleviated to some extent by the inclusion of sequences, such as insulators, or locus-control regions (Wolf et al., 2000), but it is impossible to predict whether a given construct will show the desired pattern of expression after integration. While these effects do not directly affect the safety or utility of those animals that are eventually used, they do introduce considerable inefficiency into the system

A further problem with obtaining correct expression of an introduced transgene is that introduced genes are subjected to silencing by processes including methylation of C residues at CpG dinucleotides, which frequently are found in chromosomal regions important in the regulation of gene expression. Methylation is a major mechanism for turning off the expression of inappropriate genes in somatic cells. Silencing can occur in somatic tissues, but is particularly acute with introduced genes after passage through the germline, where there is widespread DNA methylation at an early stage of embryogenesis (Jaenisch, 1997). Normal genes in their proper place have signals—in most cases unknown—that reverse the methylation at the appropriate developmental stage. Such signals generally are not present on many commonly used promoter elements, such as retroviral LTR sequences, and expression directed by these elements rarely survives passage through the germline (Pannell and Ellis, 2001). As a natural example, humans carry thousands of endogenous proviruses that have resulted from retroviral infection of the germline of our distant ancestors; yet only a very few are ever expressed at any significant level (Boeke and Stoye, 1997). Proviruses based on commonly-used murine leukemia virus (MLV) vectors introduced by deliberate infection of the germline suffer a similar fate (Jahner and Jaenisch, 1985). Recently, it has been found that vectors based on human immunodeficiency virus (HIV) can be used to efficiently insert genes into the germline of mice, and that genes inserted in this way are not subject to silencing following germline transmission (Lois et al., 2002; Pfeiffer et al., 2002). Such vectors promise to yield more efficient and reliable means of generating transgenic animals of many species. Similar vector systems based on the distantly related feline immunodeficiency virus (FIV) and bovine immunodeficiency virus (BIV) also are being developed (Curran et al., 2000; Berkowitz et al., 2001). Again, methylation-induced shutoff of gene expression is an issue affecting the strategy and efficiency of production of transgenic animals, much less their safety as producers of useful products.

Necessity for Selection

As the discussion above indicates, germline modification remains a hit-or-miss technology and, with most techniques, only a very small fraction of the progeny obtained has the desired properties of expression, copy number, and lack of genetic damage. Thus, large numbers of animals must be screened for the presence and copy number of the inserted sequence, for its properly regulated expression, for the ability of this expression to survive transmission through the germline and, finally, for the desired phenotypic characteristics and absence of unintended genetic side effects (see below and Chapter 6). Such screening could require several generations of breeding before one can be confident of the absence of recessive genetic damage, and the failure rate of the overall process is very high. As nuclear transfer technology improves, techniques requiring direct introduction of DNA into the animal germline followed by extensive screening of progeny are likely to be replaced by much simpler manipulation and selection of cells in culture, followed by recreation of animals with the desired properties directly from the nuclei of the manipulated cells (Brink et al., 2000).

CONCERNS RELATED TO GERMLINE TECHNOLOGY

There are a number of safety issues that arise as a consequence of manipulation of the germline. These can be divided into several levels of concern: from the animal (or group of animals); to the human handler, recipient, or user of the animal or its products; to the human population as a whole; and the environment. All of these levels are discussed here in the context of the technology used; many are presented in more detail in the following chapters.

As discussed above, introduction of DNA into a cell—whether somatic or germline—is not a well-controlled process and can lead to a number of undesired genetic consequences.

Unintended Genetic Side Effects

Introduction of DNA into random sites in the germline is a mutagenic event that will affect any gene that happens to be at or near the site of integration. The most obvious effect is the disruption of the integrity of a gene into which the insertion occurs. Since a large fraction of the mammalian genome is noncoding DNA derived from various kinds of silenced transposable elements, not all integration events will lead to gene inactivation; however a fraction of animals selected for the presence of a transgene has been found to carry associated genetic lesions. In mice, for example, it has been estimated that about 5 percent of MLV proviruses integrated into the germline have led to mutations o this sort (Boeke and Stoye, 1997). Direct DNA introduction can lead to numbers of integrated copies at multiple sites, leading to a risk of creating animals with a variety of genetic defects, which should be carefully screened for in the course of subsequent breeding. For example, one of the very first transgenic mouse lines generated, intended to contain an inserted active oncogene, also suffered a lesion that caused a severe recessive developmental limb defect (Woychik et al., 1985). A number of other examples of insertional inactivation by transgenes introduced into mice are known, and this approach has been proposed as a useful technique for mutagenesis (Woychik and Alagramam, 1998). Two additional points should be noted. First, the successfully transfected embryo might have inserted DNA sequences other than those that express the transgene, so the point of damage can be at a location different from the active transgene. Second, damage of this sort is often (but not always) recessive, so that it can only be detected by inbreeding to derive animals homozygous at the site(s) of the inserted DNA, adding to the complexity of the screening process.

A related effect is the activation of gene expression in the vicinity of the transgene through the action of the introduced promoter elements. This sort of inappropriate activation of expression is the mechanism of cancer induction in animals infected by a variety of retroviruses, and it has been well-studied as a model for oncogenesis. There are a number of mechanisms by which the expression of genes adjacent to (or even at some distance from) the integration site can be activated, including promoter and enhancer insertion, as well as gene fusion and introduction of elements that stabilize messenger RNA (Rosenberg and Jolicoeur, 1997). Indeed, alteration of expression of genes at genome sites far removed from a transgene has been reported in cell lines, apparently due to altered methylation (Muller et al., 2001). Whether this effect also occurs in transgenic animals is not known. Activation effects are likely to reveal themselves as dominant mutations that can have a variety of phenotypic consequences, from derailing normal development to causing a high rate of cancer later in life.

Unexpected Effects of the Modification

Even if expressed as desired, the genetic engineering itself can often have unexpected effects on the physiology of the engineered organism. One example of such an unwanted effect relates to the xenotransplantation model described above. The galactosyl transferase deficiency in humans, which leads to hyperacute rejection of organ from pigs, also is thought to offer a level of protection against zoonotic infection by enveloped viruses (Weiss, 1998). This effect occurs because the surface proteins of viruses produced by nonhuman cells are also engineered with the same galactosyl-galactose structure found on host cell proteins, and are therefore subject to the same potent immune response. This response would lead to the rapid elimination of viruses transmitted from animals before infection could occur. Pigs that are engineered by knockout of this gene would, therefore, have the potential to transmit viruses, such as influenza, much more readily to human handlers. A related concern is that human cell-surface proteins introduced into animal species as transgenes could render those animals susceptible to human viruses, increasing their risk of disease and providing alternative hosts for the spread of human disease. For example, the human poliovirus receptor (*CD155*) renders mice susceptible to poliovirus infection when introduced as a transgene (Racaniello and Ren, 1994). Also, the human complement-response modifying proteins *CD46* and *CD55*, which are being introduced into pigs to protect xenografts from rejection, also serve as receptors for human viruses—measles and Coxsackie, respectively (Weiss, 1998). Their presence in transgenic pigs not only could render these animals susceptible to infection by the human viruses, but also could provide a new evolutionary pathway for adaptation of pig viruses to human cells. Since the receptors for many other viruses have not yet been identified, the potential for this sort of surprise exists whenever a human cell-surface protein is introduced into another animal species.

Marker Genes

Vector constructs used for creating transgenic organisms usually contain genes other than the desired transgene. These genes are typically drug-resistance markers obtained from bacteria, which also can confer resistance to the same or similar drugs on eukaryotic cells. The *neo* gene encoding neomycin phosphotransferase, for example, has been used widely for selecting cells in culture infected with retrovirus or other gene constructs (vectors). In most cases, marker genes remain in vectors used for generation of transgenic (especially knockout animals). While many researchers in the field consider them a relatively harmless convenience, there is a potential for them to cause undesired side effects to the host species (such as aiding in the generation of novel antibiotic resistant pathogens) or the ultimate consumer (such as acting as novel allergens). While their potential for real harm is probably very small, it is difficult, maybe impossible, to prove that marker genes are harmless in consumer products. Such genes are usually unnecessary to the product itself and can usually be dispensed with by sound experimental design. Their presence raises concerns about the food and environmental safety of genetically engineered animal products.

Undesired Inserts

In addition to insertion of the correct element at multiple locations, the preparation of material used to generate the transgene (or knockout) might contain additional sequences unrelated (or only partially related) to the one of interest and intent. Even extensively purified DNA fragments derived from plasmids grown in *E. coli* might still contain large amounts of contaminating material derived from the host bacterium. Because such fragments can be heterogeneous in size and sequence, they are difficult to detect in DNA preparations by standard methods like gel electrophoresis.

A particular problem in this regard arises with retroviral vectors, because host cells (especially of mouse origin) often contain large numbers of endogenous virus and virus-like sequences that can, in some cases, constitute a majority of the genomes present in vector preparations (Chakraborty et al., 1994; Scadden et al., 1990). Inadvertent introduction of such sequences into the germline of transgenic animals not only has the potential for creating unintended genetic damage, but also can contribute by recombination to the generation of novel infectious viruses. A well-known example is the inadvertent generation of replication-competent MLV's containing multiple such recombinants during the growth of a vector containing a globin gene (Purcell et al., 1996). These viruses were highly pathogenic in rhesus monkeys, causing a fatal lymphoma, similar to the disease induced by MLV in mice.

Potential for Mobilization

When viral vectors are used for the introduction of genes into the germline of animals, there exists a potential for inadvertent transmission of the gene to other individuals (not necessarily of the same species). This undesirable effect could occur if such an animal were to be infected with a virus sufficiently similar to the vector to package the vector into virions. For example, if a transgenic chicken were created using an avian retrovirus vector, then infection of the transgenic chicken with any related virus (such viruses are quite commonly found in commercial poultry operations) could lead to the production and release of a virus that could transmit the gene to other animals where its presence and expression might be highly undesirable, such as among wild bird populations. Generation of a replicating virus could occur in the absence of exogenous infection, since many species contain endogenous retroviruses in their genomes that could serve as agents of this kind of mobilization. For example, in cats carrying murine leukemia virus-based vector constructs, the introduced genes could be mobilized to other cats (or, at least theoretically, to their human hosts) by the endogenous feline leukemia viruses found in most animals. As discussed above, the use of vectors based on HIV has the potential to improve the efficiency of introduction of new genes into the germline of many animal species. Such germline vectors could, in principle, also be mobilized by HIV or a sufficiently close relative. Viruses closely related to HIV are found only in African primates; however, viruses of the same genus (*Lentivirus*) are fairly common in cats (feline immunodeficiency virus or FIV), cattle (bovine immunodeficiency virus or BIV), and sheep (visna-maedi virus or VMV; Rosenberg and Jolicouer, 1997). Despite the distant relationship, FIV has been shown to transfer HIV-based vector constructs from one cell to another, raising a serious concern that similar transfer of genes introduced by an HIV (or any lentivirus) vector could be mobilized among animals infected with these common viruses (Berkowitz et al., 2001; Browning et al., 2001).

A related concern arises with the use of *mariner* and related transposons (including *sleeping beauty)* to introduce germline DNA. Related elements have been found in large numbers (14 thousand copies) in the human genome (Lander et al., 2001) and planaria, nematodes, centipedes, mites, insects (Robertson, 1997), and humans (Robertson and Zumpano, 1997), suggesting the possibility of horizontal gene flow via transposition among highly diverse hosts (Robertson and Lampe, 1995; Hartl et al., 1997; Hamada et al., 1997; Kordis and Gubensek, 1998; 1999; Jordan et al., 1999; Sundararajan et al., 1999). These potentially could be mobilized by the constructs used to transfer *mariner*-like elements into the germline, and their insertion into genes could give rise to unexpected genetic damage. Horizontal gene transfer also might be mediated by ingestion of DNA (Houck et al., 1991; Yoshiyama et al., 2001).

The possible importance of horizontal gene transfer in eukaryotes is controversial (Cummings, 1994; Capy et al., 1994); the most compelling argument for horizontal gene flow in eukaryotes is the ubiquity of transposable elements and endogenous retroviruses in genomic DNA, with no known means for their distribution other than by horizontal gene transfer.

It should be noted that any groups using transposable elements for genetic engineering could express the transposase or hopase in the trans configuration and then delete the gene for these enzymes from the transgene constructs, so that once inserted into the host's chromosome, the element is immobilized. Were this a requirement applied to transposable element vector systems for genetic engineering of animals, the hazards at issue could be minimized or eliminated, so long as active elements capable of mobilizing the introduced sequences were not already present in the host animal.

Potential for Creation of New Pathogens

In addition to their potential for mobilization by interaction with related viruses, transgene sequences also can contribute elements to infecting agents that might modify their ability to cause disease. The donation of drug-resistance genes to bacteria as a consequence of their widespread presence in transgenic livestock is one theoretical example, although the resistance gene would have to be one not found in the environment for the risk of such an event to be significantly enhanced over the natural background. Another example is the possible generation of new retroviruses following recombination between endogenous or exogenous viruses and ones used as vectors for transgenes. This recombination event could result in the provision of new genes or regulatory elements (such as LTR's capable of more efficient expression) that could adversely modify the pathogenic potential of the infecting virus. A recent natural example is the generation, through recombination between an infectious avian retrovirus and a distantly related endogenous element, of a highly virulent virus, called HPRS-103, or subgroup J avian leukemia virus (ALV) (Payne et al., 1992; Benson et al., 1998). This virus apparently arose as the result of a single, very rare event, but subsequently has been spread worldwide and has become a source of considerable economic loss to poultry breeders (Venugopal, 1999).

ISSUES RELATED TO SOMATIC CELL NUCLEAR TRANSFER TECHNOLOGY

The generation of animals using nuclear transfer from somatic cells has received a great deal of attention recently, and it is clear that this technology is fast becoming a practical way to rapidly propagate animals with valuable properties (Polejaeva et al., 2000). The DNA genomes of somatic cell nuclei used for this procedure differ in two important ways from those of germline cells. First, they have shortened telomeres at the ends of the chromosomes, a consequence of multiple rounds of cell division in the absence of telomerase, the enzyme responsible for maintenance of telomere length. Since loss of telomere length is the principal mechanism limiting the lifespan of cells in culture (Urquidi et al., 2000), the lack of appropriate-length telomeres might be expected to reduce the lifespan of the newly generated offspring or their progeny, but, surprisingly, telomere length (and lifespan of cultured cells) are restored to normal values following generation of cattle by somatic cell nuclear transfer (Betts et al., 2001), even when senescent cells are used to donate nuclei (Lanza et al., 2000). Second, the methylation state of the DNA of somatic cells is quite different from that of germline cells (Rideout et al., 2001). Since methylation (at CG sequences) plays a major role in the overall regulation of gene expression, it might be expected that inappropriate methylation states might lead to gross developmental abnormalities in embryos produced by somatic cell nuclear transfer. Indeed, it is possible that the inability of the embryo to properly reprogram methylation and expression is a major cause of the developmental abnormalities often seen in the generation of NT-produced embryos (Rideout et al., 2001). However, the apparently rapid increase in success rate of this procedure with experience, combined with the fact that animals who survive to adulthood are apparently normal (Betthauser et al., 2000), implies that correct methylation can be restored in NT embryos under the proper conditions. "Correct conditions" might involve having the transferred nucleus in the proper stage of the cell cycle (Gibbons et al., 2002), but this point is controversial. Furthermore, in a direct study (Kang et al., 2001), it was found that correct methylation and expression levels of several key genes were restored in pig embryos derived from adult cell nuclei. Thus, although nuclear reprogramming is a significant practical issue in the efficient application of this technology, it does not appear to present as insurmountable a barrier as once thought. Apparently the developmental process has a much more robust error-correction system than believed possible a few years ago.

The committee carefully considered the possible concerns that might be raised by use of somatic cell nuclear transfer technology. A few issues regarding animal welfare could be identified (see Chapter 6), including the possibility of inappropriate gene expression during development due to altered methylation patterns, or other developmental problems, such as oversized fetuses (Young et al., 1998), as well as concerns that the widespread application of this technology might reduce genetic diversity of animal populations. However, the effects of cloning are more difficult to anticipate because competing processes are at issue. On the one hand, cloning by its nature produces identical copies of a particular individual, reducing genetic variability relative to what would have been transmitted via conventional breeding. On the other hand, cloning makes it possible to save and utilize genetic variability that would not otherwise be available, for example, the genetic resources from a steer proven to be high performing. The tradeoff between the competing processes is hard to quantify in the absence of simulation modeling with validation from field observations. Whatever the mechanism causing it, loss of genetic diversity could limit the potential for future genetic improvement of breeds by selective breeding or biotechnologic approaches. Further, disease could spread through susceptible populations more rapidly than through more genetically diverse populations.

This latter concern is well documented and several studies illustrate the susceptibility of species with low genetic diversity to infectious disease. Diversity of animal populations, particularly at major histocompatibility (MHC) loci, is a major factor preventing spread of disease, particularly viral disease (Xu et al., 1993; Schook et al., 1996; Kaufman and Lamont, 1996; Lewin et al., 1999). Different MHC types recognize different viral or bacterial epitopes encoded by pathogens for presentation to the immune system. In genetically diverse populations, pathogens can evade the immune response only if they adapt to each individual MHC type following transmission from one individual to another. The requirement for this evolutionary process provides a population of animals with significant protection against the spread of infection. Pathogens can more easily evade host immune response in genetically uniform populations (Yuhki and O'Brien, 1990). The consequences of the failure of immunorecognition is illustrated by the deadly epidemics of diseases—such as measles—spread by initial contact between Europeans and isolated New World populations that lacked adequate MHC diversity. Not only could enhanced susceptibility create significant risk for the spread of "new" infectious diseases in "monocultures" of cloned or highly inbred animal populations; it also could create new reservoirs for spread of zoonotic infections—like new strains of influenza—to humans. The seriousness of these concerns, particularly relative to current practice (see Chapter 1) obviously must vary considerably from one type of animal to another, and might be alleviated with further technologic advances.

CONCLUSIONS

The technology for modifying the germline of domestic animals is being advanced at a very rapid pace. Indeed, some major advances were reported during the brief period in which this report was being prepared. Although many of the detailed issues discussed in this chapter will no doubt soon become outdated to be replaced by new ones not yet considered, some general issues will remain. In particular, there will (probably) always be concerns regarding the use of unnecessary genes in constructs used for generation of engineered animals, the use of vectors with the potential to be mobilized or to otherwise contribute sequences to related environmental organisms, and the effects of the technology on the welfare of the engineered animals themselves.

3

Animals Engineered for Human Health Purposes

INTRODUCTION

Since genetic engineering has the potential to alter the uses to which domestic animals are put, it also can lead to fundamental changes in the relationship between (1) individuals of the same species or population, (2) different species, (3) engineered animals and their products, and (4) the products and humans. There currently are major research efforts underway to develop the use of genetically engineered animals as sources for production of nontraditional materials for human use. Such uses can be divided into three major categories: biopharmaceuticals for animal or human use; live cells, tissues, and organs for xenotransplantation; and raw materials for processing into other useful end products (the latter use is discussed in Chapter 4). Several possible concerns that might in practice arise from the first two uses are discussed in the following sections.

BIOPHARMACEUTICAL PRODUCTION

A large number of genes encoding useful protein products—hormones, blood proteins, and others—has been introduced into domestic animals, leading to their expression in milk, eggs, or blood (Dove, 2000; Table 3.1). So far, none of these animals has been used for commercial production. However, a recent report suggests that the same technology might be extended to the large-scale production of vaccines (Stowers et al., 2001). Such "biopharming" applications have the potential to use well-established agricultural methods to produce large amounts of valuable products at relatively low expense as compared to fermentation. Although the end products of these applications will be novel, by and large, the process of production and the potential concerns are not likely to differ greatly from those seen in current practice, such as the use of animals or animal cell cultures to prepare live vaccines (Brown et al., 2001), hormones, or traditional products such as meat, milk, or leather. These standard products are subject to specific regulatory procedures, and essentially the same regulatory framework is expected to apply for products of both biopharming and standard technology as regards common issues such as purity of the final product, microbial contamination, levels of adventitious DNA, and the like. Nevertheless, a few more specialized concerns arise.

Contamination or Spread of Novel Pathogens

As discussed in Chapter 2, there is a theoretical potential for microorganisms to acquire—by recombination or transduction—genes from the vector constructs used to insert the transgene. Although there is no example yet of acquisition of any gene, including drug resistance markers, by bacterial flora living in a transgenic animal, the spread of introduced genes remains a possibility, albeit remote.

Of greater concern is the possibility for generation of potentially pathogenic viruses by recombination between sequences of the vector used to introduce a transgene and related, but nonpathogenic, viruses that might be present in the same animal. These concerns are particularly acute for retroviral vectors. Retroviruses appear to be efficient vehicles for inserting transgenes into many species, including chickens (Crittenden and Salter, 1990; Briskin et al., 1991), mice (Jahner et al., 1985), cattle (Chan et al., 1998), fish, and shellfish (Sarmasik et al., 2001), and might prove more successful than pronuclear injection of DNA in the generation of transgenic offspring. In many species, including chickens and pigs, there are endogenous proviruses (including the porcine endogenous viruses, PERVs, discussed below) that are competent for low-level replication in the host animal, but have no apparent pathogenic consequences (Boeke and Stoye, 1997). Endogenous proviruses are DNA sequences that were derived from infection of germline cells with a retrovirus and that are transmitted from parent to progeny like any normal gene. Their attenuation relative to their exogenous, pathogenic counterparts often is due to differences in transcriptional regulatory sequences in long terminal repeats (LTR's; Rosenberg and Jolicoeur, 1997). Since many vectors, such as the widely-used ones derived from murine leukemia virus (MLV), have LTR sequences derived from pathogenic viruses, the presence of both vector and endogenous provirus in all cells of a transgenic animal provides the potential for generating pathogenic recombinant viruses by straightforward and well-understood mechanisms. Such concerns are particularly acute in chickens and pigs, where infectious proviruses very similar in sequence to those used for vectors are known to be present (Boeke and Stoye, 1997). In mice, there is a well-studied model in which recombination between benign endogenous proviruses or endogenous proviruses, and infecting viruses early in the life of the animal, can cause a high incidence of lymphoma (nearly 100 percent in some mouse strains) 6 months later (Stoye et al., 1991; Rosenberg and Jolicoer, 1997). Given this example, it is reasonable to expect that viruses of much greater pathogenicity are likely to arise in an animal when there is a possibility of recombination between vector and endogenous viral sequences.

Similar concerns arise with the use of vectors based on lentiviruses for the introduction of genes (see Chapter 2). Recombination of lentiviruses in circulation in domestic animal populations, such as Feline Immunodeficiency Virus (FIV) in cats and Bovine Immunodeficiency Virus (BIV) in cattle, with vectors based on Human Immunodeficiency Virus (HIV) is improbable due to the large genetic distance between them. However, vectors based on FIV and BIV are being developed (Curran et al., 2000; Berkowitz et al., 2001), and their use to introduce transgenes into the corresponding species would significantly increase the probability of generating more pathogenic recombinants.

TABLE 3.1 Potential uses of transgenic animals for pharmaceutical production.

Species	Theoretical Yield (g/yr of Raw Protein)	Examples of Products Under Development
Chicken	250	Monoclonal antibodies Lysozyme Growth hormone Insulin Human serum albumin
Rabbit	20	Calcitonin Superoxide dismutase Erythropoietin Growth hormone IL-2 α-glucosidase
Goat	4,000	Antithrombin III Tissue plasminogen activator Monoclonal antibodies α-1-Antitrypsin Growth hormone
Sheep	2,500	α-1-Antitrypsin

		Factor VIII
		Factor IX
		Fibrinogen
Cow	80,000	Human serum albumin
		Lactoferrin
		α-Lactalbumin

Source: Modified from Dove, 2000.

Ensuring Confinement of Unwanted Animals

Although biopharm animals are not intended for consumption by humans or other animals, there are grounds for concern that adequate controls be in place to ensure that this does not happen without appropriate approval (see Chapter 4). As long as they do not contain the product of the introduced gene, there might be no strong reason to believe that eating or using products from transgene-containing animals would pose a threat to human health; the possibility of such a threat combined with the lack of regulatory oversight for such uses argues strongly for confinement measures.

Although it has been stated that such animals will be too valuable to the owners to allow their misappropriation (Wall, 2001), the fact that the products of interest usually are produced only by lactating females means that half the transgene-containing animals essentially will be valueless, as will the females at the end of their period of useful production. "No takes," or animals generated from manipulated embryos, but culled because of lack appropriate expression of the transgene product (or lacking the transgene itself) also are inevitably generated in significant numbers during the production of transgenics. Thus, companies using biopharm animals are likely to seek approval for marketing food or rendered products from surplus animals, and the regulatory agencies will need to be ready to deal with such requests. Of greater concern is the possibility that surplus animals (and their carcasses) might, through inadvertence or theft, find their way into the food or rendering chain, or be used for breeding, thus allowing uncontrolled spread of the transgene into the general population. This would create a regulatory problem of dealing with unapproved transgenes after their release into the food chain—a problem analogous to that posed by the appearance in food products of Starlink, a transgenic maize unapproved at the time for human consumption (Fox, 2001).

XENOTRANSPLANTATION

Xenotransplantation differs from other uses of genetically engineered animals in that it has the potential to create something entirely new—permanent human–animal chimeras—in which cells of distantly-related species survive and function for long periods of time in the most intimate contact possible. Given its potential for alleviating human diseases due to irreversible tissue or organ failure (Table 3.2), and given the acute shortage of human organs for transplant, there are very active research programs underway, in both commercial and academic laboratories, to overcome the significant immunologic and physiologic barriers, and thereby to bring xenotransplantation into standard medical practice. This topic and associated concerns about infection have been reviewed in great detail elsewhere (Boneva et al., 2001), and only an overview is given here.

At present, the only animal under serious consideration as a xenotransplant donor is the pig. For regulatory purposes, human cells cultured *ex vivo* with the cells of any other animal, such as mouse cell lines, also are considered to be xenotransplants (DHHS, 2001); co-cultivation with mouse cell lines has been used in the preparation of some cultured skin grafts as well as human stem cell lines; Thomson et al., 1998). While nonhuman primates, such as the baboon, would seem to have physiologic and immunogenetic advantages such as the lack of a hyperacute immune response, their scarcity as well as the difficulty of clearing them of adventitious infectious agents (as well as ethical concerns) render them impractical for further consideration.

The field of xenotransplantation covers a great many procedures, ranging from implantation of single cells to treat Parkinson's disease and tissues, such as pancreatic islets, to treat diabetes; extracorporeal use of intact organs, such as perfusion of patient blood through pig livers to provide short-term support in cases of liver failure; to transplantation of whole organs—heart, kidney, liver, and so on. While whole-organ xenotransplantation remains far in the future, development of the simpler modalities is underway, and hundreds of human

TABLE 3.2 Applications of xenotransplantation.

Indication	Transplant	Status
Organ Failure	Pig heart, kidney, liver, etc.	0
Acute liver failure	Extracorporeal perfusion	1
Diabetes	Pancreatic islets (or cells)	1
Parkinson's disease, Huntington's Disease, Focal epilepsy, Stroke	Neural tissue	1
Burn, Skin injury	Skin autograft (co-cultured with mouse cells)	2

Note: 0=No successful experience.
1=Some trials have been performed.
2=Successful trials have been performed.

subjects have received porcine cells or tissues as part of clinical trials in the United States, Russia, Israel, and many European countries (Paradis et al., 1999). Given the nature of infectious disease issues, regulatory concerns are not limited to the United States alone, but extend to the international health community as well.

The development of xenotransplantation as a part of clinical practice promises great benefits in terms of making possible essentially infinite supplies of replacement tissues and organs where severe shortages exist today. This development naturally will entail both great potential benefit as well as considerable risk to the study participant, but such risk is not qualitatively different from that entailed in the development of any other new medical procedure and will not be considered further. The principal concern is that the uniquely close relationship created between recipient and host will allow novel opportunities for transmission of infectious disease, and possible creation of new disease agents in the process. While the history of close contact between humans and pigs is a very long one, and one would imagine that all possible transmission of infectious agents between the two species already would have been seen and thoroughly studied, it is possible that the "co-culture" environment of a transplant would be qualitatively different in ways that would allow different outcomes. Two different types of agents are discussed separately.

Exogenous Infectious Agents

In general, bacteria and parasites that might cause problems readily can be excluded from source flocks, leaving viruses as the principal concern (Onions et al., 2000). As can be seen in Table 3.3, the number of viral agents that are of potential concern is very large. Not all of the viruses are on the list because of their potential to cause human disease; some would cause serious disease among the donor animals and others are sensitive indicators of breaks in biosecurity, and so forth. In principle, since all of theses agents are horizontally (one animal to another) or vertically (mother to offspring) transmitted, they can be eliminated by proper management—proper containment, vaccination, close monitoring, culling, birth by Caesarian section, etc. In practice, elimination is going to prove a very difficult task, since the numbers of agents are very large and there is a lack of reliable assays for detecting many of them. Nevertheless, problems resulting from transmission of exogenous infectious agents are not qualitatively different from the present situation with human donors (allotransplantation), where infection with agents transmitted with the transplanted organ (such as Epstein–Barr virus and cytomegalovirus) is a major problem. In fact, it is anticipated that reduction in the risk of acute morbidity

TABLE 3.3 Exogenous pig viruses of concern in xenotransplantation.

Family	Species	Category
Picornaviridae	Foot and mouth disease	
	Enterovirus 1 Talfan/Teschen	2, 5
	Enterovirus (other serogroups)	5
	Enterovirus swine vesicular disease	5
	Human enteroviruses	1
	Encephalomyocarditis	
	Rhinovirus	5
Caliciviridae	Enteric calicivirus	1
	Swine hepatitis E	1
Astroviridae	Porcine astrovirus	5
Togaviridae	Western encephalitis	1
	Eastern encephalitis	1
	Venezuelan encephalitis	1
	Getah	1
	Chikungunya	1
Flaviviridae	Japanese B encephalitis	1
	Louping Ill/TBE complex	1
	Wesslebron disease	1
	Apoi	2
	Dengue fever	1
	West Nile fever	1
	Classical swine fever (hog cholera)	5
	Bovine viral diarrhoea	5
	Border disease	5
Coronaviridae	Transmissible gastroenteritis	4, 5
	Porcine respiratory coronavirus	4, 5
	Epidemic diarrhea	4, 5
	Haemagglutinating encephalomyelitis	4, 5
	Porcine reproductive & respiratory disease syndrome	4, 5
	Porcine torovirus	5
Paramyxoviridae	Murine parainfluenza virus type 1 (Sendai)	2
	Parainfluenza 2	2*
	Parainfluenza 3	2
	Blue eye disease	5
	Menangle	1
	Nipah	1
Rhabdoviridae	Vesicular stomatitis	1
	Rabies	1
Bornaviridae	Bornavirus	2, 5
Orthomyxoviridae	Influenza A	1
	Influenza C	5
Bunyaviridae	Akabane	1, 5

	Batai	1, 5
	Hantavirus	1, 5
Arenaviridae	Lymphocytic choriomeningitis	1, 5
Reoviridae	Ibaraki	5
	Reovirus 1 to 3	2
	Rotavirus A, B, C, E.	2
Birnaviridae	Porcine picobirnavirus	5
Retroviridae	Porcine endogenous	2
Hepadnaviridae	Hepatitis B	*
Circoviridae	Porcine circovirus	5
Parvoviridae	Porcine parvovirus	4, 5
Papovaviridae	Porcine polyomavirus	3
	Porcine genital papillomavirus	3, 5
Adenoviridae	Porcine adenovirus serotypes 1 to 4	3
Herpesviridae	Pseudorabies	2
	Porcine cytomeglovirus	5
	Porcine lymphotropic herpesvirus type 1	3
	Porcine lymphotropic herpesvirus type 2	3
Poxviridae	Swinepox	5
	Vaccinia	2
	Cowpox	1, 5*
	Orf/pseudocowpox	1, 5*
Desoxyviridae	African swine fever	5

NOTE: 1=Zoonotic.
         2=Replicates in human cells or weak evidence for zoonotic potential.
         3=Might undergo abortive replication and possibly oncogenic replication.
         4=Belongs to a family with evidence of frequent changes in host range or pathogenicity.
         5=Undesirable as indicates a breakdown in biosecurity and/or might compromise health of
           the pigs.
*=Although the virus has not been detected in pigs, it has been included for reasons such as its wide
host range.
Source: Onions et al., 2000. Courtesy of D. Onions.

and mortality resulting from the transmission of infectious agents with
transplanted organs will be a significant benefit of xenotransplantation.

Porcine Endogenous Retroviruses

PERVs present quite a different situation and level of concern since they are inherited as part of the host genome and, therefore, cannot be removed easily from donor animals. All pigs contain multiple (around 50) PERV proviruses in their genome, at least several of which encode infectious virus. PERVs are gammaretroviruses, closely related to MLV, that can be classified into three subtypes, A, B, and C, based on their envelope gene sequences (Takeuchi et al., 1998). Subtypes A and B can infect many types of human cells in culture. Subtype C is much less infectious for humans. Most breeds of pig carry proviruses capable of yielding infectious virus of all three subtypes. Although most pigs carry about the same number of proviruses in their DNA, there is considerable diversity in location, implying that their insertion into the genome must have occurred rather recently (on an evolutionary time scale). Based on extensive experience with related endogenous proviruses of mice, it is highly likely that the majority of proviruses contain some sort of genetic defect, and that only a small number are responsible for release of infectious virus. Taken together with the polymorphism in the presence or absence of specific proviruses, it might well be possible to breed animals lacking infectious proviruses for use as xenotransplant donors.

PERVs have not yet been shown to cause disease (or even viremia) in pigs or any other species in which they have been detected. Nor has their presence been detected (by polymerase chain reaction, PCR, or serology) in more than 150 human recipients of pig cells or tissues (Paradis et al., 1999), although a low level of infection of recipient cells can be observed in immunodeficient mice transplanted with porcine islets of Langerhans (Van der Laan et al., 2000). Nevertheless, given the release of viruses infectious to human cells by many types of pig cells; the close similarity of these viruses to viruses known to cause cancer, immunodeficiency, and other diseases in mice and cats; the well-known adaptability and variability of retroviruses; and the example of the rapid worldwide spread of HIV and AIDS, there is serious concern that the novel association between pig and human tissues might create novel evolutionary opportunities for the virus, leading to the appearance of a new pathogen. Although such a pathogen could have serious long-term adverse consequences for the transplant recipient, this issue is not an area of concern since it is far outweighed by the potential benefits of the transplant. The real issue of concern is that the xenotransplant setting might provide the opportunity for the virus to evolve into a pathogen that also could be transmitted from one individual to another efficiently enough to create a new epidemic disease.

Such an evolutionary pathway would require a series of events, each increasingly improbable, as indicated by the scale shown in Table 3.4 (J. P. Stoye, 2001). As implied by the table, it is virtually certain that many cells in the transplant would express infectious PERV following transplantation, and it is likely that some local infection of host cells would occur. The subsequent events necessary for generation of pathogenic, transmissible viruses increasingly are unlikely, but on some unknown, arbitrary scale. Although the probability of inadvertent creation of a new epidemic generally is judged to be extremely small (particularly given the long history of intimate association between humans and pigs), it cannot be ignored altogether. Current FDA policy is to permit xenotransplantation trials to proceed, but to require close monitoring of recipients, and (insofar as possible) of their contacts (DHHS, 2001). Attempts also are being made to identify specific proviruses responsible for production of infectious virus and then to selectively breed them out of lines of animals to be used as transplant donors (Herring et al., 2001).

TABLE 3.4 Theoretical scale of risks associated with PERV transmission from xenotransplants.

Event	Cumulative Probability
Expression of infectious virus	High
Localized infection of host cells	
Spreading infection in the host	
Persistent viremia	
Disease (e.g., lymphoma, "AIDS")	
Transmission to close contacts	
Spreading, epidemic transmission	Extremely low

4

Food Safety Concerns

SCOPE AND GOAL

This chapter identifies potential food safety concerns for meat or animal products derived from animal biotechnology. The species considered include beef and dairy cattle, sheep, goats, poultry, swine, rabbits, and a wide array of finfishes and shellfishes.

The scope of this chapter encompasses: (1) non-genetically engineered animals that are propagated by nuclear transfer or other cloning techniques, (2) genetically engineered animals developed primarily for meat, milk, or eggs, and (3) genetically engineered animals developed for biomedical or industrial products. This latter category is considered because entry of these animals into the food chain might be proposed at the end of their productive life or sooner, as in the case of unused females and males, which typically are not used to generate the recombinant product (e.g., bulls in which the recombinant protein is expressed in the mammary gland).

The criteria used for identifying important scientific issues were developed considering the hazard (i.e., a compound or agent that has the potential to produce harm), the likelihood of harm resulting from exposure to the hazardous compound or agent, the likelihood that exposure to the hazard would occur, and the severity of any harm that would be realized. In this context, harm ranges from allergic reactions to other forms of illness, including, in the extreme case, death. Concerns are described on a scale ranging from no concern, to low level of concern, to moderate level of concern, and to high level of concern.

BACKGROUND

Interest in the quality and safety of food of animal origin began to develop in the United States in the latter part of the nineteenth century, aimed primarily at meat for export to Europe. Regulatory and inspection systems for domestically produced red meat (but not poultry, eggs, or milk) were initiated in 1906 (Wiser, 1986). The U.S. Food and Drug Administration (FDA) prepared a list of food safety hazards (Foster, 1982) about 20 years ago. Food-borne toxigenic and pathogenic microorganisms were named first and considered to be the greatest danger to consumers. Also included were malnutrition, environmental contaminants, toxic natural constituents, reaction products that are formed during processing or preparation and storage for eating, pesticide residues, and finally, food additives. Food safety concerns raised by the use of animal biotechnology add to this list.

As new threats to food safety were recognized, new technologies and regulatory protocols were developed to enhance the safety of food. The occurrence of parasites was managed with slaughterhouse inspections, new husbandry systems, and parasiticidal drugs (Hagstad and Hubbert, 1981). By 1978, only 12 food-borne cases of parasite infection were documented (U.S. Public Health Service, 1978). Residues from drugs used to improve animal health and productivity arose as a food safety concern, but monitoring and inspection protocols largely have been effective in preventing illegal or unsafe levels of residues in food (Meyerholz, 1983; NRC, 1999; FDA, 2000). Microbial pathogens originating in animal fecal material remain the primary concern for the safety of food of animal origin (Tauxe, 1997).

Microbes pathogenic to humans grow in the animal gastrointestinal (GI) tract, and might or might not cause health problems in the animal (Altekruse et al., 1997). Physiologic stress increases the susceptibility of the animal to pathogens, the growth of pathogens in the GI tract, and their shedding into the fecal material of the stressed animal (Salminen et al., 1998). These same pathogens might enter the human food chain when they are transferred to the surface of the meat during slaughter and processing. The role of human food safety related to pathogens from animal fecal material was fully recognized only in the latter part of the twentieth century (Tauxe, 1997).

Secondary concerns for food safety arise from the disposition of carcass remains after removal of the edible meat, and from the disposal of animal fecal material. (Potential environmental concerns related to fecal material from genetically engineered animals are discussed in Chapter 5.) After the edible meat is removed, carcass remains are processed into other products used in a variety of applications, including food and medical uses (Klinkenborg, 2001). One of the major products is meat and bone meal (MBM), a supplement historically fed to high-production animals. Using MBM from infected cattle in animal feed can transfer bovine spongiform encephalopathy (BSE) to other ruminants, and ultimately, to human consumers, which has occurred in Europe but not in the United States (Bruce et al., 1997). Concern about BSE transmission in the United States has resulted in regulations forbidding the feeding of MBM to ruminants (FDA, 1997). Animal carcasses also are used in a number of other products. Collagen is processed into gelatin for confectionary products such as candies, capsules for pharmaceutical products, and a range of cosmetic products. Bone and connective tissues are used in bone grafts and hernia repair in humans. Therefore, concern for the safety of products derived from animals also must take into account the use to which carcass remnants might be put once the edible portions are removed.

FOOD PRODUCTS FROM NON-GENETICALLY ENGINEERED CLONED ANIMALS

The cloning technologies of embryo splitting (EMS; Willadsen, 1979; Williams et al., 1984) and blastomere nuclear transfer (BNT; Willadsen, 1986; Prather et al., 1987) using embryo cells were introduced into dairy cattle breeding in the 1980s (Chapter 1). Although not widely adopted, a total of 1,472 EMS cloned Holstein females was registered with the American Holstein Association through 2001 (Norman et al., 2002) and evaluated genetically for yield traits, meaning they produced calves and were milked commercially. Yields of female EMS clones were greater than those of the Holstein population by 189 kilograms (kg) milk, 8 kg fat, and 7 kg protein, but slightly less than those of noncloned full siblings. The latter result might indicate an impact of the technology on performance or slightly different management of the two groups. Of 754 EMS cloned bulls registered and 143 evaluated by the U.S. Department of Agriculture (USDA) as sources of donor sperm, only 22 had noncloned full siblings. Results of the evaluations of the sires are not yet available. A total of 187 BNT cloned Holsteins (61 males and 126 females) were registered through 2001 (Norman et al., 2002); 74 had milk yield records, but only 11 had noncloned full siblings. The yields and milk composition of BNT clones exceeded those of the national herd average by 278 kg milk, 10 kg fat, and 10 kg protein, but were similar to those of their noncloned full siblings.

Although existing data for EMS and BNT clones addresses the changes in milk yield and composition, they do not specifically address the food safety of their milk and meat products. Aside from a study on yearling Brangus bulls that compared body measurements and measures of carcass merit obtained from their steer clone-mates (Diles et.al., 1996), there are no published analytical studies of meat and milk composition comparing the products of cloned animals and full siblings evaluating in detail any unanticipated compositional differences, differences in protein quality, or nutrient bioavailability.

Since the donor nuclei used to produce EMS clones are taken from embryonic cells, there is little if any genomic reprogramming needed to drive embryogenesis. However, blastomeres from embryos of more than eight cells, (i.e., from the stages typically used for BNT), must be reprogrammed upon NT (Van Stekelenberg-Hamers et al., 1995; Kono, 1997; Bordignon et al., 2001), since they express a substantial number of genes, including paternal genes, that are not expressed by the oocyte nucleus. Indeed, the nucleus of the mature donor oocyte is transcriptionally quiescent and is associated with a different set of chromatin proteins (e.g., histones) compared to the recipient oocyte nucleus. A similar array of gestational and postnatal abnormalities seen in somatic cell nuclear transfer also has been observed in BNT (Wilson et al., 1995; Garry et al., 1996; Wilmut et al., 1997; Cibelli et al., 1998) clones. To the degree that inadequate or otherwise different reprogramming relative to that occurring normally in gametic nuclei occurs in BNT (De Sousa et al., 1999; Daniels et al., 2000), the composition of food products from NT animals might differ from that of ordinary animals. Although it is difficult to characterize the level of concern without specific data, it seems unlikely that there are changes in gene expression directly related to EMS and BNT cloning procedures that would raise nutritional or food safety concerns. Food products from BNT clones have been consumed by humans, with no apparent ill effects. Based on current scientific understanding, the committee regards products of EMS and BNT clones as posing a low level of food safety concern. Nevertheless, it would seem appropriate that the FDA use available analytic tests to evaluate the composition of food products from animals that themselves result directly from BNT cloning procedures to verify that they fulfill existing standards for animal-derived food products. The products from the offspring of cloned animals were regarded as posing no food safety concern because the animals are the result of natural matings.

The cloning of animals from somatic cells is more recent. Limited sample size and health and production data, as well as rapidly changing cloning protocols, make it difficult to draw conclusions regarding the safety of milk, meat, or other products from somatic cell clones and their offspring. The key scientific issue is whether and to what degree the genomic reprogramming that occurs when a differentiated nucleus is placed into an enucleated egg and forced to drive the development of a clone might result in gene expression that raises food safety concerns. Differences in patterns of developmental gene expression in non-engineered individuals and somatic cell clones would be greatest during early development when reprogramming is incomplete. A number of datasets suggest that the health and wellbeing of neonatal and young somatic cell clones often are impaired relative to those of normal individuals (see Chapter 6 regarding animal wellbeing). Direct effects of any abnormalities in patterns of gene expression on food safety are unknown. However, because stress from these developmental problems might result in shedding of pathogens in fecal material, resulting in a higher load of undesirable microbes on the carcass, the food safety of products, such as veal, from young somatic cell cloned animals, might indirectly present a food safety concern. As a somatic cell clone develops and nuclear reprogramming is completed, patterns of gene expression would approach those of a non-engineered individual. Indeed, the health and wellbeing of somatic cell clones approximated those of normal individuals as they advance into the juvenile stage. For example, somatic cell cloned cattle reportedly were physiologically, immunologically, and behaviorally normal, and exhibited puberty at the expected age, with high rates of conception upon artificial insemination (Lanza et al., 2001). Two of these individuals have given birth to calves that seem phenotypically normal. There are to date no published comparative analytical data assessing the composition of meat and milk products of somatic cell clones, their offspring, and conventionally bred individuals (although several studies are in progress; Bishop, personal communication, 2002). However, the committee found it difficult to characterize the level of concern without further supporting evidence regarding food product composition. Currently, there is no evidence that food products derived from adult somatic cell clones or their progeny pose a hazard (i.e., there is no evidence that they present a food safety concern).

GENETICALLY ENGINEERED ANIMALS

A number of types of genetically engineered animals will be developed primarily for food, and others will be developed primarily for producing non-food materials such as pharmaceuticals, vaccines, fibers, and other high value products. The principles for assessing the safety of food from genetically engineered animals are qualitatively the same as for non-engineered animals, but animals genetically engineered for non-food products might present additional concerns relating to the nature of the products that they generate. As for all foods or food products, those from genetically engineered animals should be evaluated for agents—chemical or biologic—which affect the safety of the food for the human consumer.

Animals used for xenotransplantation are not considered safe for human consumption and are excluded from the food chain by current regulations (see Chapter 7 for information on food animal regulations). Their exclusion is based primarily on concerns about persistent tissue residues of agents used to anaesthetize the animal prior to harvesting the tissues and organs. If there were any possibility that such animals might be rendered and considered for further processing into useful human food or medical products, concerns about anesthetic residues would remain pertinent. If animals genetically engineered for xenotransplantation, but not used for that purpose, were presented for entry into the food chain, the food safety of such animals also would have to be evaluated based on protocols developed for evaluating other genetically engineered animals.

Animals might be genetically engineered to produce non-food products in their milk or eggs. Half of the genetically engineered population will be male, and will not be directly useful for production of heterologous proteins in, for example, milk or eggs. It is likely that companies producing such animals will seek early entry of males that are transgenic, but incapable of producing milk or eggs, into the food chain. In addition, companies might want to enter females that are "no takes", which do not express high levels of the product of interest, or that have reached the end of their productive lives, into the human food supply. The safety of food products from such animals that were culled from transgenic lines might present concerns.

Numerous experiments have shown that the level and specificity of transgene expression in an animal is predictable only to a limited extent, probably because all the factors affecting gene expression have not yet been identified (Houdebine, 2000). Transgenes might be expressed at a low level in various tissues in which the promoter is not expected to be active. Such ectopic expression might be due to genomic position effects attributable to the action of neighboring enhancer elements. In addition, ectopic expression might result from basal-level transcription at the site of integration (Ashe et al., 1997; Travers, 1999). Recombinant proteins whose expression is driven by regulatory elements directing expression in mammary glands have been observed in the blood of transgenic animals during lactation (Bishoff et al., 1992; Devinoy et al., 1994; Thepot et al., 1995). The presence of transgene products in blood might result from leakage of the mammary epithelium or from secretion at the apical side of mammary cells. For example, although the promoter from the whey acidic protein (WAP) gene has been used to direct expression of a transgene in mammary tissue, and some concentration of WAP normally is found in the blood of lactating animals (Grabowski et al., 1991). Hence, through bioactivity, allergenicity, or toxicity pathways, ectopic gene expression might directly affect the safety of food products derived from tissues, sexes, or life stages of transgenic animals where transgene expression is not expected. In some cases, recombinant proteins produced in milk have deleterious effects on mammary gland function (Bishoff et al., 1992; Shamay et al., 1992; Bleck et al., 1995; Ebert et al., 1994) or on the transgenic animals more generally (Burdon et al., 1991; Reddy et al., 1991; Jhappan et al., 1993; Devinoy et al., 1994; Hennighausen et al., 1994; Thepot et al., 1995; Massoud et al., 1996; Litscher et al., 1999). These effects might stem from ectopic expression of the transgene or from transfer of the recombinant proteins from mammary gland to blood. Animals with variable levels or ectopic expression of the transgene presumably will be identified in the development of the transgenic lines. Should products from such individuals be released to commercialization channels, they could pose a food safety concern unless the protein of concern is screened for and found absent. It is expected that well-established transgenic lines to be used in routine production will have been subjected to selection, and that concerns posed by unstable or ectopic gene expression will have been addressed to a large degree. Should pharmaceuticals or other biologically active proteins enter the food supply through products of such animals, associated food safety concern could be high. Additionally, the effects of transgene expression on animal wellbeing might indirectly affect the safety of food products derived from their tissues through stress-mediated mechanisms.

Expression of transgenes also might be intended to change the nutritional attributes or improve the safety of food products. For example, expression of transgenes in milk might optimize milk composition, add neutraceuticals to milk, or reduce the incidence of infectious disease (Zuelke, 1998; Houdebine, 2000). Several systems are being developed to reduce lactose concentration in milk (Alton et al., 1998; Whitelaw, 1999). Secretion of bovine α-lactalbumin (an enzyme) in pig milk increased piglet growth (Bleck et al., 1998; Wheeler, 1994), showing the potential for changing the nutritive value of milk. Immunoglobulin A directed against viruses infecting the digestive tract might be expressed in milk (Saif and Wheeler, 1998; Castilla et al., 1998; Sola et al., 1998), and viral antigens activated by oral administration might be used to vaccinate humans and animals against viral disease (Houdebine, 2000). Changes of these types raise a moderate level of food safety concern. Claims of nutritional attributes, safety, and efficacy of milk or other food products from transgenic animals must be demonstrated.

Animals might be developed to produce food products designed to fit special human dietary needs. Possible future products might include milk that lacks the most common allergenic protein, eggs that are lower in cholesterol, meat with enhanced vitamin content, or fat content modified in quality or quantity (Young, 2002). The nutrient profiles of meat and animal products are well documented, and changes in this profile raise concerns. Changes might be unwanted by some consumers, and might add value for others. If these changed products were labeled in order to appeal to targeted consumers, and identifiable to those who have medical or other reasons to avoid such foods, they would be of low concern. Novel proteins also can be produced by genetic engineering. Although proteins are necessary components of the human diet, they can exert undesirable effects, including: (1) allergenicity and hypersensitivity, (2) bioactivity, and (3) toxicity.

Allergenicity and Hypersensitivity

Food allergies are adverse reactions to a protein or glycoprotein in food that elicits a heightened response of the immune system in some people. Among several types of immunologic responses causing food allergies, the most common type of reaction is mediated by allergen-specific immunoglobulin E (IgE) antibodies. IgE-mediated reactions are known as immediate or acute hypersensitivity reactions because symptoms occur within minutes to several hours after ingestion of the allergenic food. Food allergies also include delayed hypersensitivity reactions whose mechanisms are less clear. These include cell-mediated reactions where the onset of symptoms occurs more than eight hours after ingestion of the allergenic food. In the United States, the prevalence of food allergies is 1.5 percent of the general population, and 5 percent of children under three years of age (Sampson, 1997). The prevalence of these types of reactions in infants remains uncertain, but cases have been well documented (FAO, 2001). Many children outgrow their food allergies (Sampson, 1997; Taylor et al., 1999). There are eight foods or food groups that account for more than 90 percent of the food allergies in the United States. These include cow's milk, eggs, fish, crustaceans, peanuts, soybeans, tree nuts, and wheat (Taylor et al., 1999). However, more than 160 other foods have been identified as causing food allergies (Hefle et al., 1996).

The genetic engineering of animals intended for use as food will involve the expression of new proteins in animals; hence the safety, including the potential allergenicity of the newly introduced proteins, will have to be assessed. While most known allergens are proteins, only a few of the innumerable proteins found in foods are allergenic under typical circumstances of exposure (Taylor and Hefle, 2001). While the common sources of food allergens have been identified and characterized, many others are less known and poorly understood. If the new protein originates from a known allergenic source or its amino acid sequence is similar to that of a known allergen, the protein can be tested to determine whether it causes a reaction with sera from individuals with known food allergies. However, the potential allergenicity of a protein can be reasonably assessed *only* when the protein is known to trigger an immune response in sensitive subjects. By contrast, the potential allergenicity of a protein of unknown allergenicity cannot easily be predicted, as no immunosera of allergic subjects are available (Mendieta et al., 1997). A more difficult issue arises when a new protein comes from a source that historically is not a human food. Assessing the potential allergenicity of transferred proteins remains one of the most difficult aspects in the overall safety assessment of transgenic foods. An adequate allergenicity assessment will require an understanding of several factors, including the source of the transferred protein, its level of expression, the physical and chemical properties of the protein, and any structural similarities to known allergens. No single factor can be considered definitive, but consideration of all these factors together might provide some indication of potential allergenicity (Gendel, 1998 a,b; Taylor and Hefle, 2001). Concerns regarding the potential allergenicity of these new compounds in food are the lack of predictive and testing methodologies, and the feasibility of performing adequate assessments for an increasing number of transgenic products. The possibility that particular novel gene products might trigger allergenicity or hypersensitivity responses will vary with the gene product at issue, and poses a moderate level of food safety concern (i.e., the likelihood of a reaction is of moderate concern, but when it occurs, it could be a severe). The committee notes that the World Health Organization and other bodies are working to develop and standardize protocols for testing allergenicity.

Other Bioactive Compounds

In some cases, the aim of genetic engineering is to enhance expression of an economically important trait (e.g., growth rate; Pursel et al., 1990; Devlin et al., 2001), or to improve resistance to disease (e.g., mastitis; Kerr et al., 2001). In others, animals will be engineered to express proteins of pharmaceutical interest (Wright et al., 1991). These applications involve the expression of biologically active proteins or polypeptides encoded by a transgene. The possibility exists that such molecules could retain their bioactivity after consumption, raising a food safety concern.

The bioactive product of a transgene, in most cases, will be a protein or, in some cases, a polypeptide. During digestion, proteins and larger polypeptides largely are broken down into small peptide fragments and amino acids by proteolytic enzymes in the digestive tract. Di- and tripeptides that are absorbed into digestive epithelial cells are broken down into amino acids by intracellular enzymes. Few intact, small peptides are absorbed into the bloodstream during digestion. Substantial degradation of the intact protein effectively destroys its original bioactivity, since that bioactivity depends on the integrity of at least a portion of the protein or peptide. Allergenicity might, of course, remain a problem for sensitive individuals. Many food allergens are absorbed into the bloodstream. While most intact proteins generally are not absorbed into the bloodstream of healthy adults with an intact, properly functioning digestive system, absorption might occur in individuals whose digestive epithelium has been compromised by disease or injury (Simon, 1985), possibly posing allergenic response. In such cases, the normal array of digestive enzymes might be absent, or the integrity of the epithelium as a barrier might be compromised. Gastroenteritis, for example, can reduce the secretion of digestive enzymes and cause the breakdown of digestive epithelium, resulting in the passage of intact proteins and peptides into the bloodstream. A food safety concern thus arises when individuals whose digestive system has been compromised by disease, injury, or advanced age ingest foods containing bioactive proteins or peptides.

The digestive epithelium of newborn infants permits the transient absorption of whole proteins or large protein fragments until closure of the gut epithelium occurs. Closure is facilitated by breastfeeding and delayed in infants that are formula fed. The timing of the closure might range from weeks to months, depending on dietary factors. Prior to closure, a wide variety of intact proteins might cross the digestive epithelium by a non-selective mechanism and enter the bloodstream. Thus, consumption of food (especially milk) containing bioactive proteins or peptides could result in the transfer of such molecules into the bloodstream of newborn infants. This possibility raises a concern regarding recombinant bioactive molecules present in milk used in infant formulas.

Bioactive peptides and proteins also might exert their effects in the digestive system, prior to absorption. For example, recombinant human bile-salt stimulated lipase (BSSL) has been expressed in the milk of transgenic sheep; this protein is intended for oral administration as a therapeutic agent for treating patients suffering from pancreatitis (PPL Therapeutics, 2001). Consumption of food products (i.e., milk and meat) from animals expressing bioactive molecules such as BSSL could alter digestion in otherwise healthy individuals, and presents a food safety concern. Lysostaphin, a bactericidal protein expressed by certain bacteria, has been expressed in murine milk, where it reduced mastitis caused by *Staphylococcus aureus* (Kerr et al., 2001). Transgenic cattle expressing lysostaphin in milk have been generated with the intent of reducing mastitis in that species (Suszkiw, 2001). Similarly, Jia et al. (2000) proposed production of transgenic fish expressing the hybrid antimicrobial peptide cecropin-melittin for control of fish pathogens. Preliminary studies using injections demonstrated the effectiveness of the antimicrobial peptide to protect fish against infections and suggested that the strategy of overexpressing the peptides in transgenic fish might provide a method of decreasing bacterial disease problems in fish. Milk containing lysostaphin or fish expressing cecropin-melittin could alter the balance of digestive tract flora of consumers of these products; in addition, widespread use of such antimicrobial agents also could foster the emergence of lysostaphin-resistant strains of pathogenic *S. aureus* or *Vibrio anguillarum*. Thus, food products containing antimicrobial proteins might present a food safety concern in view of their potential to alter the balance of consumers' intestinal flora, and might foster the evolution of microbial strains resistant to specific agents.

Many genetically engineered fish and shellfish express an introduced growth hormone (GH) gene—most often a fish GH gene—in order to promote rapid growth. Hence, it is particularly important to make sure that such a transgene product has no biologic activity in humans or animals that consume fish or shellfish expressing such a transgene. The food safety of GH proteins was evaluated when administration of recombinant bovine GH (rbGH, also called somatotropin) to dairy cattle was considered by the FDA in the late 1980s (Juskevich and Guyer, 1990). The FDA cited data showing that non-primate GH proteins are not biologically active in humans; nor are fragments of the GH molecule, nor insulin-like growth factors secreted by the host in response to GH administration. Neither bovine GH nor bovine insulin-like growth factor I (IGF-I) were orally active in rats, a species responsive to parenterally administered bovine GH. The FDA also cited studies showing that bovine, ovine, whale, and porcine growth hormones are not biologically active in humans, which suggests that piscine growth hormones are unlikely to be biologically active in humans. The degree to which full-term human infants absorb intact proteins is equivocal; FDA cited studies showing that concentrations of IGF-I in milk of rbGH-treated cows was within the physiologic range in human breast milk, and IGF-I is denatured under conditions used to process cow's milk for infant formula. In the unlikely case that products of GH-transgenic fish or shellfish would be fed to human infants, cooking would denature active GH and IGF-I molecules.

Against the background of the discussion above, the committee regards the likelihood that a bioactive product poses a hazard will vary among gene products, food products, and consumers, in various cases posing a low to moderate level of food safety concern. For a susceptible individual, however, such a hazard could have severe consequences.

Toxicity

Many toxins are well studied and genes for known toxins would not be transferred purposefully into a food animal. As noted earlier, genetic engineering will have the potential of introducing novel proteins expressed in food animals. Because proteins generally are broken down in the digestive system into common amino acids, the direct toxicity of proteins (beyond the possibilities of allergenicity and bioactivity discussed above) is unusual and generally of low food safety concern. Purposefully expressed proteins that remain intact or otherwise pose a potential safety concern presumably will be fully evaluated in the pre-market review and approval process, and thus pose a relatively low concern.

Of greater concern to the committee with respect to possible toxicity are the unintended and unanticipated effects and byproducts of the genetic engineering of a food animal, including but not limited to these novel proteins. For example, the engineering could alter a metabolic process that then results in a toxic metabolite being present in edible tissue. The question, as suggested earlier in the chapter, is whether edible products of genetically engineered animals have been screened adequately to detect the presence of unanticipated compositional changes that might introduce toxicity. Assuming that adequate analytic methods and screening protocols exist (an issue that the committee did not examine during its deliberations) and are applied, the possibility of such toxicity poses a low level of food safety concern to the committee.

5

Environmental Concerns

Potential impacts on the environment from the escape or release of genetically engineered organisms was the committee's greatest science-based concerns associated with animal biotechnology, in large part due to the uncertainty inherent in identifying environmental problems early on and the difficulty of remediation once a problem has been identified. The intent of this chapter is to identify the risks to the environment posed by GE animals, prioritize those risks, and explain the criteria used for selecting them. The committee based its assessment on principles of risk analysis that are general in their application and not limited to currently developed biotechnology. Where possible, examples from the scientific literature are used, while in others hypothetical examples are used to illustrate risks that exist in theory but thus far have not been observed.

The committee explicitly recognized that along with potential risks, there might be many benefits of biotechnology for alleviating human suffering and for addressing problems with growing food demands. The ultimate decision of when or where to use biotechnology will be evaluated not only in relation to these benefits, but also to those of alternative technologies. However, the charge to this committee was not to examine the benefits of biotechnology, or of the technical alternatives, but rather to "develop a consensus listing of risk issues in the food safety, animal safety, and environmental safety areas for various animal biotechnology product categories." The committee also was asked "to provide criteria for selection of those risk issues considered most important that need to be addressed or managed for the various product categories." By using definitions of risk and hazard established in previous National Research Council reports, the committee attempted to rank those concerns. In these two ways, the committee attempted to put those concerns in perspective and to provide a balanced viewpoint.

Any analysis of GE organisms and their potential impact on the environment needs to distinguish between organisms engineered for deliberate release and those that are engineered with the intention of confinement but escape or are inadvertently released. The discussion in this report focuses primarily on the latter category, but the committee recognized the possibility of intentional release of GE organisms into the environment and expressed a high level of concern about it. This chapter also focuses primarily on risks as a result of genetically engineered (GE) animals entering natural environments and transgene spread through vertical gene transmission (the sexual transfer of genetic information between genomes) followed by natural selection. The risk of horizontal gene transfer (the nonsexual transfer of genetic information between genomes; Kidwell, 1993) is discussed primarily in Chapter 2.

This chapter, therefore, is organized into a discussion of: (1) general principles of risk analysis, (2) general aspects of the organism, transgene, or transgene function that can be used *a priori* to prioritize GE animals for level of environmental concern, (3) risks posed by key classes of GE animals, and (4) the need for further research directed at improving our understanding of hazards and estimating risks posed by genetically engineered animals.

GENERAL PRINCIPLES OF RISK ANALYSIS

Consideration of environmental concerns posed by GE animals must be based on an understanding of key concepts underlying the science and practice of ecologic risk assessment. A seminal review of risk assessment methodology (NRC, 1983) states, "Regulatory actions are based on two distinct elements, risk assessment, and risk management. Risk assessment is the use of the factual base to define the health effect of exposure of individuals or populations to hazardous material and situations." Risk management is "the process of weighing policy alternatives and selecting the most appropriate regulatory action, integrating the results of risk assessment with engineering data and with social, economic, and political concerns to reach a decision." Clearly, risk management is beyond the purview of this committee, while elements of risk assessment are needed to prioritize concerns.

Understanding Risk: Informing Decisions in a Democratic Society (NRC, 1996) updated the 1983 NRC study and provided two important definitions: *Hazard*: an act or phenomenon that has the potential to produce harm, and *Risk*: the likelihood of harm resulting from exposure to the hazard. While the earlier study describes risk assessment as containing some or all of the following steps:

(1) hazard identification, (2) dose-response assessment, (3) exposure assessment, and (4) risk characterization. These steps do not apply well to GE organisms in the environment because dose-response and exposure assessments are intended to apply to substances that can be quantified in discrete amounts and that cannot reproduce themselves. Adapting principles from both studies (NRC, 1983; 1996) to the current problem, the committee used the definitions of risk and hazard to develop a set of working steps.

Defining Risk

Risk, as defined, is a probability that can be quantified and expressed in an equation, thereby providing a method to prioritize concerns. However, exact probabilities of risk might be difficult or impossible to determine for all categories of possible harm. Indeed, all possible harms might not be known or knowable *a priori*, particularly with respect to secondary effects (see Chapter 7 for a discussion of unknown harms). On the other hand, based on current knowledge of population genetics and receiving ecosystems, and experience with domesticated species, it is possible to classify GE organisms into categories of high to low probabilities of spread into the environment. Risk of possible harms (known and unknown) can then be inferred from the probability of spread (i.e., risk of harm to a healthy natural population is low), if the transgene is purged from the population. This method is used only to prioritize the likelihood of a GE organism to destabilize a natural community; it does not address possible harms to humans, direct or indirect (direct risks to human health are examined in Chapter 4).

Because risk is the joint result of exposure and harm, it is the product of two probabilities: the probability of exposure, $P(E)$, and the conditional probability of harm given that exposure has occurred, $P(H|E)$, that is, Risk, $R = P(E) \times P(H|E)$. In this context, the steps in risk analysis are: (1) to identify the potential harms regardless of likelihood, (2) to identify the potential hazards that might produce those harms, (3) to define what exposure means for a GE organism and the likelihood of exposure, $P(E)$, (4) to quantify the likelihood of harm given that exposure has occurred, $P(H|E)$, and (5) to multiply the resulting probabilities to prioritize risk. Because all potential harms might not be known or cannot be known (see Chapter 7), it will be necessary to update this procedure continually as knowledge accumulates, using an adaptive management approach (NRC, 1996; Kapuscinski, 2002).

PRIORITIZING GE ANIMALS FOR LEVEL OF ENVIRONMENTAL CONCERN

Steps in Ecologic Risk Assessment

Identifying Potential Harms and Hazards

In an ecologic context, harm is defined as gene pool, species, or community perturbation resulting in negative impacts to community stability. These include displacement or reduction in the number of species that exist in a community or numbers within each species. This definition is all-encompassing and broad, but can be further refined once a particular GE organism is identified and the environment into which it might escape or be released is known.

The hazard is the GE organism itself because it is the agent that might cause negative impacts to community stability. These negative impacts might be either direct (e.g., resulting from direct competition for limited food or resources)—or indirect, caused by changes in other biotic factors utilized or needed by the ecologic community (Scientists' Working Group on Biosafety, 1998).

The process of prioritizing concerns will vary from case to case because of the uniqueness of each GE construct, transgenic founder individual from which a line is derived, and receiving ecosystem (USDA, 1995). However, based on the principles of risk assessment, the committee attempted to prioritize environmental concerns posed by GE animals by considering the following variables: (1) the effect of the transgene on the "fitness" of the animal within the ecosystem into which it is released, (2) the ability of the GE animal to escape and disperse into diverse communities, and (3) the stability and resiliency of the receiving community. These three variables determine the likelihood that a GE organism will become established in a receiving community—a critical factor in risk assessment.

Defining What Exposure Means for a GE Organism and the Likelihood of Exposure: $P(E)$

Exposure is a threshold phenomenon because an initial escape or release of a GE organism might not have a measurable effect on the receiving community; the organism might not be able to establish itself in the community, and might be lost rapidly due to natural selection. Thus, provided the natural population is not already endangered, exposure must be more than just release or escape for a GE organism to prove a hazard. The GE organism must spread into the community. The committee, therefore, defines exposure as the establishment of a GE organism in the community, and in the following text, establishment will be substituted for exposure. For risk assessment, the critical factor is the likelihood the GE organism will become established in a community, which is *P(E)*. This conclusion does not mean that risk cannot occur without establishment. As discussed later, if a transgene causes local species extinctions, either because the population size is critical or because the transgene produces a Trojan gene effect, considerable harm might result. However, these are special cases that can be addressed as such. The likelihood of establishment is dependent on an organism's fitness and ability to escape and disperse in diverse communities (Scientists' Working Group on Biosafety, 1998), and the qualities of the receiving community.

Fitness

Once a transgene is introduced into a community, whether by vertical or horizontal gene transfer, natural selection for fitness will determine the ultimate fate of the transgene if the population is large enough to withstand the initial perturbations (Muir and Howard, 2001). Fitness is quantified relative to that of other individuals in the population and is simply the genetic contribution by an individual's descendants to future generations of a population (Ricklefs, 1990). Fitness in this context refers not only to its survival component, but also its reproductive component, that is, to *all* aspects of the organism's phenotype that affect spread of the transgene. Muir and Howard, in modeling the potential spread of a transgene (2001; 2002a,b), reduced these aspects to six net fitness components: juvenile and adult viability, age at sexual maturity, female fecundity, male fertility, and mating success. The model is based on the assumption that natural selection acting through these components will determine the ultimate fate of the transgene.

The last component, mating success, often is overlooked because it generally is not a factor in artificial breeding programs; it often is, however, the strongest factor driving natural selection (Hoekstra et al., 2001). For example, increased adult size in most species of fish is positively correlated with mating success (as, for example, in many salmonid species: Jones, 1959; Schroder, 1982; Jarvi, 1990; Groot and Margolis, 1991). With Japanese medaka (*Oryzias latipes*), males 25 percent above average in size realized a 400 percent increase in mating success (Howard et al., 1998). Such increases in mating success could result in the spread of a transgene even if the transgene reduces survival rate (Muir and Howard, 1999).

From a population genetics perspective, if a GE organism is more fit than its wild relatives in the receiving population, the GE organism eventually will replace its relatives or become established in that community. If it is less fit, the engineered trait eventually will be removed from the receiving population. If the fitness of transgenic and nontransgenic individuals is similar, the likely outcome is persistence of both transgenic and nontransgenic genotypes (Hedrick, 2001; Muir and Howard, 2001).

The effect of genetic engineering on fitness can be determined either prospectively or retrospectively. Appendix A of the Scientists' Working Group on Biosafety (1998) provides a prospective assessment of factors that would affect an organism's ability to become established in the environment, while Muir and Howard (2001a,b; 2002) provide a retrospective method based on measurement of net fitness components.

From a prospective view, the key factor affecting fitness is transgene functionality within the GE organism. Functionality can be divided into four broad categories: those that increase adaptability of the GE organism to a wider range of environmental conditions, usually through new functionality; those that alter existing traits for improved performance within standard production agriculture; those that produce new or novel products; and those that produce animals or animal products for human medical benefit.

Increased Adaptability

A transgene might increase an organism's adaptation to a wider range of environmental conditions, for example, by increasing freeze tolerance (Fletcher et al., 1992) or removing a limiting growth factor, perhaps allowing the organism to synthesize an amino acid that was previously limiting, or to digest previously indigestible carbon sources such as cellulose, or to obtain phosphorous from previously inaccessible sources, such as phytic acid (Golovan et al., 2001a,b). Finally, a transgene can be used to increase disease resistance by, for example, disabling retroviruses, producing coat proteins that activate the immune system against certain viruses or that bind to receptor molecules by which viruses enter cells, or by producing antibiotics to protect against bacterial infections (Dunham et al., 2002; Jia et al., 2000; Sarmasik et al., 2002).

Such adaptations also could allow GE animals to invade or persist in ecosystems where they otherwise could not, such as salt or brackish water, while maintaining populations in communities where they normally occur, such as freshwater lakes and streams. Such a combination could result in a sustained invasion of the new community from the species' original or introduced range until complete colonization results. Hence, a transgene that increases fitness or adaptation increases the probability of establishment and results in the highest level of concern for establishment.

Enhanced Existing Traits

Selective breeding as well as genetic engineering have enhanced the productivity and growth of many domesticated farm animals. Many transgenic animals have been engineered for enhanced growth rates (Hammer et al., 1985; Pursel et al., 1987; Devlin et al., 1994; 1995a; 1995b; Rahman and Maclean, 1999). Production traits in domesticated farm animals include, for example, growth rate, feed efficiency, egg number, milk yield, litter size, and fiber yield (e.g., wool). Experience with conventional selection for such traits in domesticated farm animals suggests that such modifications do not increase the fitness of animals in natural environments, often because of physiologic imbalances or growth demands in excess of the food available in natural environments. Transgenic animals designed to meet these objectives might be even less fit than those developed using selective breeding.

Selective breeding is based on manipulation of polygenic inheritance, in which the resulting phenotype results from the cumulative effect of changes in allele frequencies of many genes with a distribution of effects from small to large (Lynch and Walsh, 1998) and which are selected over multiple generations. In contrast, transgenesis involves one or few genes with relatively large effects, introduced in a single founder generation. In the selective breeding process, the correlated traits needed to support enhanced growth and reproduction, such as skeletal and vascular systems, also are selected for indirectly; this is not always the case with transgenics (Farrell et al., 1997; Muir and Howard, 2001; 2002b; see Chapter 6 regarding animal wellbeing concerns). Because of these homeostatic imbalances, domesticated animals transgenic for enhanced production traits might exhibit a greater reduction in fitness than their selectively bred counterparts. Experience with GE animals developed to date tends to support this contention/notion. For example, swine transgenic for growth hormone displayed a number of fitness problems (see Chapter 4). Similarly, fish transgenic for growth hormone have a reduced juvenile viability (Dunham, 1994; 1996; Muir and Howard, 2001; Devlin et al., 2001). Collectively, these findings seem to indicate that GE organisms developed for production traits have a low probability of establishment.

However, environmental concerns posed by animals expressing these types of transgenes cannot be dismissed. First, it is possible for GE organisms to overcome viability disadvantages if other fitness components are enhanced, such as mating success, fecundity, or age at sexual maturity (Muir and Howard, 2002b). Second, the introgression of genes decreasing fitness poses a near-term demographic risk to small receiving populations (i.e., small populations might not remain viable until the transgene is selected out, which poses a risk if a threatened or endangered or otherwise valued population is at issue). Finally, the magnitude of phenotypic change that is possible with transgenesis could exceed that of conventional breeding or natural mutations. Transgenic organisms can be produced with changes in physiologic traits far beyond what is possible with naturally occurring mutations such as dwarfism or gigantism in mammals and poultry. These naturally occurring mutations are limited to approximately four times the size of a normal organism, while, for example, transgenic salmonids have been reported to grow to a mean size-at-age of four to eleven times normal (e.g., Devlin et al., 1994; 2001).

At the heart of the issue is how species evolve. Domestication is widely believed to be the consequence of small incremental changes in trait value, and the ecologic niche of the animal is not changed if the phenotype of a mutant individual is only slightly changed. Expression of transgenes, however, could cause mega-mutations that instantaneously and substantially change the phenotype of the transgenic organism. In terms of evolutionary theory, such a mega-mutation could give rise to a switch from the currently-occupied adaptive peak to another peak on the adaptive topography of Sewall Wright's (1969; 1982) shifting balance theory. If such a shift were to occur, the GE organism might be able to establish itself in a new community or to shift its niche within the current community. An illustrative example of a natural major mutation causing a shift in evolutionary trajectory was a major mutation for mimicry that occurred in the evolution of butterflies (Lande, 1983). The primary predator avoidance attributes in butterflies are to remain concealed (crypsis) or to resemble closely another species that is distasteful to predators (mimicry). Intermediate individuals that are neither effectively cryptic nor good mimics are likely to be eaten, thus selection acting by small steps cannot account for such evolutionary adaptations. Therefore, natural mutations followed by selection can and do result in new evolutionary lines. Similarly, the expression of a growth hormone transgene producing up to 17.3-fold greater difference in weight by 14 months of age in trout (Devlin et al., 2001) acts as a mega-mutation that, for example, could change an organism from being a prey of one species to being a predator upon it.

Establishment of domesticated animals in the environment as a result of adaptive peak shifts, either through conventional or transgenic technology, has not been documented. Hence, the concern for this mode of transgene establishment in natural populations is moderate to low based on currently available evidence. However, it is theoretically possible for organisms engineered for production traits to become established in communities as a result of adaptive peak shifts; any such establishment would pose a high level of concern.

Production of New or Novel Products

Animals that are genetically engineered to produce new or novel products are yet another example of transgene functionality that could influence fitness.

Milk, egg white, blood, urine, seminal plasma, and silkworm cocoons from transgenic animals are candidates to produce recombinant proteins on an industrial scale (Houdebine, 2000). Animals also can be used to produce pharmaceuticals in eggs (Harvey et al., 2002) or milk (Wright et al., 1991), or fibers such as spider silk in milk (Kaplan, 2002). Such alterations in physiology will result in additional energy demands without conferring any obvious fitness advantage. Such transgenic animals might have little chance of establishment in the environment (excepting silkworms), and hence raise the lowest levels of environmental concern. However, other indirect aspects of expressing such products are still a concern, and will be discussed in a following section.

Production of Animals or Animal Products for Human Health and Medical Benefits

Three categories of animals are genetically engineered for human health and medical benefits: pets altered to reduce allergens, animals altered for xenotransplantation purposes (Tearle et al., 1996; Lai et al., 2002), and insects altered to control the spread of pests and diseases (Braig and Yan, 2002; Spielman et al., 2002). The first two categories most likely will either not change fitness or will result in a decline in fitness and, like animals engineered to produce new or novel products, raise the lowest levels of concern with respect to the animal's ability to establish itself in natural communities. The last category—insects altered to control the spread of pests and diseases—has mostly involved the modification of mosquitoes not to carry parasites, and has unknown effects on the fitness of the mosquito. Some reports indicate that the parasite load reduces the fitness of mosquitoes carrying it (Braig and Yan, 2002; Spielman et al., 2002), suggesting that transgenes decreasing the parasite load might increase fitness. In addition, changes in the insect's driver mechanisms (meiotic drive and incompatibility systems) are being proposed as a way of establishing the GE mosquito in the community. Because establishment is the objective and is critical for biocontrol using these techniques, this category of genetic engineering raises the highest probability for establishment.

Ability to Escape, Disperse, and Become Feral

Another aspect of evaluating the probability of establishment of a GE animal in a community is the organism's ability to escape, disperse, and become feral in diverse ecologic communities. This mainly is a function of the animal being transformed, though the receiving ecosystem also might be a factor (USDA, 1995).

The dispersal ability of GE animals is not known, but reasonably can be assessed from knowledge of similar domesticated species (Scientists' Working Group on Biosafety, 1998). Table 5.1 summarizes these characteristics for commonly farmed and laboratory species. Some communities in Australia and New Zealand have been affected dramatically, particularly by the rabbit, while in the United States and Europe, pigs, cats, mice and rats, and fish and shellfish have caused the greatest disruptions.

The more domesticated a species, the less likely it is to survive in natural environments. Highly domesticated species such as poultry or dairy cattle are not well adapted to natural conditions and might not be able to survive and reproduce in a natural setting. However, if wild or feral populations exist locally, the escaped transgenic organisms could breed with those and spread the transgene into populations that otherwise are well adapted to the local environment. If the GE animal is released into an area where a native wild or feral population of the same species exists, mates might be readily available, and the transgene could spread via mating. Even in areas where the GE species does not exist, it might breed with members of a closely related species with which it is reproductively compatible (e.g., transgenic rainbow trout, *Oncorhynchus mykiss*, with native cutthroat trout, *O. clarki*; see reviews of hybridization, e.g., Dangel, 1973; Schwartz, 1981; Campton, 1987).

In the North American agricultural system, certain agricultural animals are well confined. However, cattle and sheep roam open ranges in the West, feral pigs exist in Arkansas, Hawaii, Florida, and California, and range chickens and turkeys exist in many states. Extensive damage has been reported for feral insects imported to improve agricultural production, such as the gypsy moth (Lepidoptera: Lymantriidae), a species imported for use as a silkworm (Gerardi and Grimm, 1979), and the Africanized honeybee (*Apis mellifera scutellata*), a species imported to improve the foraging ability of European honeybees (Caron, 2002).

The committee concluded that animals that become feral easily, are highly mobile, and have caused extensive community damage pose the greatest concern. These include mice and rats, fish and shellfish, and insects. Animals that become feral easily, have moderate mobility, and have caused extensive damage to ecologic communities are next. These include cats, pigs, and goats. Animals that are less mobile, but have been known to become feral with moderate community impact, pose the next level of concern. These include dogs, horses, and rabbits. Finally, less mobile and highly domesticated animals that do not become feral easily, such as domestic chickens, cattle, and sheep, present the least concern.

TABLE 5.1 Factors contributing to level of concern for species transformed.

Animal	Factor Contributing to Concern					Level of Concern[6]
	Number of Citations[1]	Ability to Become Feral[2]	Likelihood of Escape Captivity[3]	Mobility[4]	Community Disruptions Reported[5]	
Insects[8]	1804	High	High	High	Many	High
Fish[7]	186	High	High	High	Many	↑
Mice/ Rats	53	High	High	High	Many	
Cat	160	High	High	Moderate	Many	
Pig	155	High	Moderate	Low	Many	
Goat	88	High	Moderate	Moderate	Some	
Horse	93	High	Moderate	High	Few	
Rabbit	8	High	Moderate	Moderate	Few	
Mink	16	High	High	Moderate	None	
Dog	11	Moderate	Moderate	Moderate	Few	↓
Chicken	11	Low	Moderate	Moderate	None	
Sheep	27	Low	Low	Low	Few	Low
Cattle	16	Low	Low	Low	None	

[1] Number of scientific papers dealing with feral animals of this species.
[2] Based on number of feral populations reported.
[3] Based on ability of organism to evade confinement measures by flying, digging, swimming, or jumping ability for any of the life stages.
[4] Relative dispersal distance by walking, running, flying, swimming, or hitchhiking in trucks, trains, boats, etc.
[5] Based on worldwide citations reporting community damage and extent of damage.
[6] A ranking based on the four contributing factors.
[7] Did not include shellfish, some of which (such as zebra mussel and asiatic clam) have proven highly invasive.
[8] Limited to gypsy moth and Africanized honeybee.

The Likelihood of Harm Given that Exposure has Occurred: *P(H/E)*

The stability and resilience of the receiving community is another factor that influences whether transitory or long-term harm results from the introduction of GE animals. Colonization by GE animals might result in local displacement of a conspecific population, which could have a disruptive effect on other species in a community, for example, by releasing competing species from resource competition or prey species from predation (Kapuscinski and Hallerman, 1990); additionally, the survival of predatory species that depend on the eliminated species could be threatened. This concern is best exemplified by the classic experiment of Paine (1966) in the rocky intertidal zone. By experimentally removing the top predator, a starfish (*Piaster* sp.), the number of species in the plot was reduced from 15 to eight. Another example is the impact of pigs on plant species diversity reported by Hone (2002). Ground rooting of feral pigs in Namadgi National Park, Australia, decreased the number of plant species, which declined to zero with intensive pig rooting. Thus, expansion of a species into new ecosystems can have a cascading impact on other species in the community with unpredictable harms (see Chapter 7 for further discussion).

Transgenes that increase fitness or adaptability also could have negative ecologic impacts if they spread into pest populations.   For example, phosphorous is an element essential for growth of all life forms. Securing this vital nutrient from the environment is critical for population growth. Phosphorous is contained within all seeds in the form of phytic acid. However, phytic acid is not digestible by non-ruminants (Golovan et al, 2001a). The addition of a phytase gene would allow GE non-ruminants such as pigs (Golovan et al, 2001b) or mice (Golovan et al, 2001a) to obtain needed phosphorus from seeds and grains, which would increase their ability to grow and produce more offspring, thereby resulting in a greater pest potential for feral pigs (Vtorov, 1993; Hone, 2002) and mice (Krebs et al., 1995; King et al., 1996).

Pleiotropic effects of transgenes that have antagonistic effects on different net fitness components can result in unexpected harms, ranging up to local extinction of the species into which the transgene is introduced (Muir and Howard, 1999; Hedrick, 2000). For example, the transgene might increase one component of fitness, such as juvenile or adult viability, but reduce another, such as fertility (Kempthorne and Pollak, 1970; Hedrick, 2000; Muir and Howard, 2002b). The effect of a transgene in this category parallels the use of sterile males to eradicate screwworms, except that in the case of sterile males they must be released continually to achieve control; a transgene that increases the viability component of fitness will spread on its own, while the reduced fertility brings about extinction, albeit over a longer time period. Fish transgenic for production of cecropins might represent a class of GE organisms that fit into this category.   Survival among channel catfish increased from 14.8% in the nontransgenic control to 40.7% fish expressing cecropins (Dunham et. al. 2002). However, pleiotropic effects on fertility were not measured. Cecropins, like some other antimicrobial products, might negatively impact survival of sperm and reduce fertility (Anderson et al, 2002; Zaneveld et al 2002). Similarly, if a transgene enhances mating success while reducing juvenile viability, less fit individuals obtain the majority of the matings, while the resulting transgenic offspring do not survive as well as nontransgenic genotypes. The result is a gradual spiraling down of population size until eventually both wild-type and transgenic genotypes become locally extinct (Muir and Howard, 1999; Hedrick, 2000). This is an example of harm as a result of a transgene that spreads into the receiving community but fails to become established because the population becomes extinct. Results of Devlin et al. (2001) suggest that transgenic fish might have this potential. They showed that rainbow trout transgenic for growth hormone were both larger at sexual maturity and lower in viability than their wild-type siblings. Although the mating success of transgenic males relative to wild-type males is presently unknown in rainbow trout, large body size is known to enhance male mating success in many salmonid species (Jones, 1959; Schroder, 1982; Jarvi, 1990; Groot and Margolis, 1991).

The conclusion that natural selection will determine the ultimate fate of a transgene assumes that population sizes of the native and/or competing populations are large enough to be able to rebound from a temporary inflow of possibly maladapted genes or competitors, thereby allowing time for natural selection to operate. Escape of domesticated animals, whether or not transgenic, into wild or feral populations also might affect wild-type populations adversely by introducing alleles or allele combinations that are poorly adapted to natural environments (Hindar et al., 1991; Lynch and O'Hely, 2001; Utter, 2002). If the wild population is sufficiently large, these alleles eventually should be eliminated by natural selection, although it might take many generations to reach selective equilibrium. Stochastic events could fix the alleles in small populations and result in extinction of those populations (Lynch and O'Hely, 2001)

Released animals also could introduce diseases or compete with native species for limited resources, causing population declines. If introduced males are sterile, but still mate with wild females, the reproductive efforts of those females are wasted, also contributing to population decline. In these regards, escaped transgenic organisms raise many of the same concerns as newly introduced species (Regal, 1986; Tiedje et al., 1989).

Finally, use of genetically engineered animals could harm the environment indirectly by changing demand for feed, number of animals used, or amount of resulting waste, and by the effects of wastes containing novel gene products on microbial and insect ecologies. Most biopharmed animals will be highly valuable and most likely will be carefully confined, but there is some likelihood that the gene products themselves would pose environmental harms. Should the milk from transgenic livestock be spilled, most novel proteins would degrade rapidly along with other milk proteins. However, not all novel proteins will degrade quickly, such as spider silk—a protein that could be expressed in milk (Kaplan, 2002). The possibility that novel proteins are present in significant amounts in the meat, stools, urine, or other secretions of the animal would need to be evaluated. Risk assessment of these products can follow traditional methods.

Long-term and transitory environmental harms are dependent on the stability and resilience of the receiving community. A community is deemed stable if and only if ecologic structure and function variables return to the initial equilibrium following perturbation from it. The community is deemed to have local stability if such a return applies for small perturbations, and global stability if it bounces back from all possible perturbations (Pimm, 1984). Resilience is the property of how fast the structures or function variables return to their equilibrium following a perturbation (Pimm, 1984). The quantitative stability of many systems has been investigated by Jefferies (1974), and mathematical methods to quantify stability were summarized by Ricklefs (1990).

These definitions potentially allow a prioritization of potential harms from GE animals based in part on the receiving community's stability and resilience. Those that are most stable will result in the least harm, with the greatest harm occurring to unstable (fragile) communities. The committee recognizes that characterization of community stability and resilience might not prove straightforward. Ricklefs (1990) states that ecologists disagree on exactly how to parameterize models used to simulate risks and predict outcomes, and that "we are far from resolving some of these questions, and the ultimate resolution, if it is possible, will likely come from reconciling a combination of viewpoints that, at present, focus separately on dynamical control, energetics, and adaptations of individual species."

Another limitation of this approach is that one cannot necessarily limit spread of a GE organism to a particular community. Thus, based on the principles of risk, one must assume the GE animal will become established in all possible communities for which it can gain access. If any one of these communities is fragile, concern for this ecosystem would be high. For this reason, the precautionary principle suggests that risk always should be assessed and managed for the most vulnerable ecosystem into which the escaped or released GE animal is likely to gain access following a given application.

Ranking the overall concerns then can be based on the product of the three variables cited above: fitness of the GE organism, its ability to escape and disperse, and the stability of the receiving community. Because the overall concern is a product of these three variables (and not the sum), if the risk associated with any one of the variables is negligible, the overall concern would be low (but not negligible). A transgene that increases the fitness of a highly mobile species that becomes feral easily raises the greatest level of concern, (e.g., a transgene conferring salt tolerance on catfish or the phytase gene in mice). A transgene that does not increase fitness in a low-mobility species that does not become feral easily raises the least concern (e.g., a gene for spider silk in cows; Kaplan, 2002). The committee stressed that these are *a priori* listings of concerns. When an actual transgenic organism is produced, for any GE animal that has the potential to become feral, those concerns can be assessed more directly by use of the net fitness approach, as suggested by Muir and Howard (2002a,b).

RISKS POSED BY KEY CLASSES OF GE ANIMALS

Examination of the Current State of Understanding, Regulatory Issues, and Key Findings Related to Hazard Assessment

Against the background of the discussion of principles of hazards, associated risks, and potential harms posed by genetically engineered animals generally, this section examines risks posed by key classes of genetically engineered animals: terrestrial vertebrates (laboratory and domesticated animals), terrestrial invertebrates (insects, mites, and other arthropods), and aquatic animals (fish and shellfish).

Terrestrial Vertebrates

The dangers of some terrestrial animals escaping and establishing themselves in the environment are considerable. Escaped cats, rabbits, mice, rats, pigs, dogs, fox, pigs, and goats have become feral and resulted in environmental disruptions in Australia, New Zealand, parts of Europe, and the western and southern United States. Any of these animals transgenic for functions that allow greater or wider adaptation to environmental conditions can pose significant ecologic harm. Such functions include, for example, increased nutrient utilization, or new metabolic pathways allowing nutrient synthesis ability, viral or bacterial resistance in any species, and heat or cold tolerance. Few GE terrestrial vertebrates have been produced that fit this category; the best examples to date are the phytase mouse and pig (Golovan et al., 2001a,b). Further studies will be needed to examine environmental implications of these and other GE terrestrial animals should they be produced.

Terrestrial Invertebrates

Insects can be genetically engineered to control the spread of pests and diseases and for other beneficial purposes. However, a number of scientific uncertainties regarding environmental harms and associated risks need to be resolved before the release of GE arthropods can be undertaken purposefully.

One of the primary alternatives to the use of insecticides for control of insects is the use of agriculturally beneficial insects, such as predators and parasitoids. Unfortunately, such beneficial insects often are destroyed by insecticide applications, yet if one waits for the beneficial insects to multiply in order to control the pest, unacceptable levels of damage to the crops already would have occurred. To address this problem, insects used for biocontrol could be genetically engineered for resistance to insecticides, thereby allowing simultaneous use of both biologic control mechanisms (Braig and Yan, 2002).

Another means of biocontrol is the release of sterile males. Unfortunately, such programs are expensive and might require the release of sterile females where the insects cannot be sexed before release. Techniques used to induce sterility, such as irradiation, often render the insect noncompetitive as a potential mate. A possible solution to these problems is to genetically engineer the insect to allow either genetic sexing, for example, through a female lethal gene, or through direct production of sterile males. Finally, GE insects can be developed to produce visual markers, such as green fluorescent protein (GFP), to determine the effectiveness of sterile release programs (Braig and Yan, 2002).

Another application of transgenesis is to control transmission of diseases by such vector organisms as mosquitoes. With GE technology, it might be possible to disrupt an insect's ability to carry and transmit diseases such as *Plasmodium*, the malaria parasite (Braig and Yan, 2002; Spielman et al., 2002; Ito et al., 2002). An environmental concern is presented because the parasite has a negative effect on the fitness of the mosquito (Braig and Yan, 2002; Spielman et al., 2002). Elimination of the parasite could result in the release of mosquitoes from a form of biocontrol, with a possible associated increase in mosquito populations. An increase in mosquitoes also could lead to increased spread of other mosquito-borne diseases to both animals and humans.

The development of molecular methods for genetic engineering of terrestrial arthropods (reviewed by Atkinson et al., 2001; Handler, 2001) has not been matched by advances in understanding how to deploy GE arthropods in practical pest management, or of how to evaluate potential harms associated with their release into the environment (Spielman, 1994; Hoy, 1995; 2000; Ashburner et al., 1998). Key issues pertaining to environmental risk (Hoy, 2000) include the possibility that transgenic insects released into the environment would pose unknown ecologic impacts, and that gene constructs inserted into insects could be transferred horizontally through known or unknown mechanisms to other species, thereby creating new pests.

If a genetically engineered arthropod is to be released within a practical pest management program, any potential ecologic risks associated with its release into the environment must be assessed, although guidelines for conducting such an assessment do not yet exist (Hoy, 1992a; 1992b; 1995). Anticipation of ecologic risks will depend upon predictions of the impact of changed abundance or dynamics of the engineered species upon resources or species with which the organism interacts in the environment, including predators, prey, competitors, and hosts.

Further, the methods by which horizontal gene transfer (Chapter 2) could occur should be investigated so that it can be determined whether and how to assess this particular hazard (Hoy, 2000). Should horizontal transfer of a transgene be demonstrated, it poses significant effects for the evolution of a species, introducing otherwise unavailable genetic material to the genome of a species (Droge et al., 1998). Horizontal gene transfer would pose no harm if the gene that is moved were lost, inactivated, or benign. However, if horizontal gene transfer confers increased fitness, perhaps by establishing the dominant, selectable antibiotic or pesticide resistance trait used in the production of the transgenic arthropod, then harm could be realized. Risk posed is not dependent solely on the frequency of transfer. Even rare events might cause ecologic impacts if the transferred gene increases the fitness of the recipient (Droge et al., 1998).

Considerable progress has been made in the development of methods for genetic engineering of the mosquito germ line and in identification of parasite-inhibiting molecules (Beernsten et al., 2000; Blair et al., 2000). Despite the technical progress, there remain important scientific questions that must be addressed prior to a program releasing GE mosquitoes (Braig and Yan, 2002; Spielman et al., 2002). Can parasite-inhibiting gene constructs indeed spread and become fixed in wild mosquito populations? In order to do so, a driver mechanism will have to be developed that would cause a disproportionate frequency of offspring of the released mosquitoes to carry the introduced construct (Braig and Yan, 2002; Spielman et al., 2002). Such driver mechanisms might include competitive displacement, meiotic drive (Sandler and Novitski, 1957), biased gene conversion, and others (Braig and Yan, 2002). The fate of parasite-inhibiting genes would be determined not only by the mechanism used to drive the fixation of the genes, but also by the magnitude of any loss of fitness in the host, and also by a range of ecologic and abiotic environmental factors. Possible human health effects posed by genetic engineering of disease vector insects are discussed in Chapter 3.

In the context of environmental concerns posed by GE arthropods, it is clear that purposeful release of transgenic arthropods will depend upon prior risk assessment and risk management. Hoy (1997) called for effective containment of transgenic arthropods in the laboratory and thorough peer review by scientists and regulatory agencies prior to any field release. However, there are no U.S. or international guidelines for containment of transgenic arthropods. Additionally, there are no proven techniques for retrieving transgenic insects after environmental release should they perform in unexpected ways.

Fish and Shellfish

Considerable research effort has been devoted to development of GE fish and shellfish stocks, as they pose considerable benefits to producers (Chapter 1). Production of some GE fish or shellfish could result in environmental benefits. For example, expression of growth hormone transgenes has been shown to increase feed conversion efficiency (Cook et al., 2000; Fletcher et al., 2000), decreasing the amount of feed needed to bring a fish to market size, while reducing wastes per unit of mass produced. Production of fish expressing a phytase transgene might allow use of less fish meal in feeds while decreasing phosphorus in effluent from aquaculture operations. However, transgenic fish and shellfish might pose environmental hazards (Kapuscinski and Hallerman, 1990; 1991; Hallerman and Kapuscinski, 1992a,b; 1993; Muir and Howard, 1999; 2001; 2002a,b). Below, the committee briefly reviews a series of empirical studies to examine potential ecologic risks posed by escaped or released transgenic fish and shellfish.

As indicated in Table 5.1, there are a number of important factors that contribute to risk. The risk factors for establishment in a community were high for all categories because: (1) cultured fish and shellfish stocks are not far removed from the wild type, (2) aquaculture production systems frequently are located in ecosystems containing wild or feral populations of conspecifics, (3) aquatic organisms exhibit great dispersal ability, and (4) aquacultured organisms often are marketed live.

Transgenic Atlantic salmon pose a near-term regulatory issue. A brief review of the hazards they pose provides a useful illustration of the environmental hazards posed by GE aquatic species more generally. Cultivated salmon escape from fish farms in large numbers (Carr et al., 1997; Youngson et al., 1997; Fisk and Lund, 1999; Volpe et al., 2000), posing ecologic and genetic risks to native salmon stocks (Hansen et al., 1991; Hindar et al., 1991). Several studies that have focused on Atlantic salmon (*Salmo salar*) expressing a growth hormone (GH) gene construct suggest that transgenesis might affect fitness, but do not provide net fitness estimations needed for parameterizing fitness models predicting outcomes should such fish enter natural systems. GH transgenic salmon consumed food and oxygen at more rapid rates than control salmon (Stevens et al., 1998); although gill surface area was 1.24 times that in control salmon, it did not compensate for the 1.6-time elevation in oxygen uptake, and the metabolic cost of swimming was 1.4 times that for control salmon (Stevens and Sutterlin, 1999). Growth-enhanced transgenic fish were significantly more willing to risk exposure to a predator in order to gain access to food (Abrahams and Sutterlin, 1999), but reduced their exposure to predators when risk was heightened further, suggesting that they might not be significantly more susceptible to predation. Transgenic salmon lost their juvenile parr markings sooner than nontransgenics, suggesting early readiness for adaptation to seawater. Thus, findings to date are fragmentary, and it is difficult to assess the likely ecologic or genetic outcome should transgenic Atlantic salmon escape captivity and invade wild populations.

Pacific salmonids include a number of aquaculturally important species that have been the subject of a large number of transgenesis experiments and a small number of risk assessment experiments. These studies collectively show results similar to those obtained with Atlantic salmon, but also show that the outcomes of introgression of a transgene might differ among receiving populations. Coho salmon (*Oncorhynchus kisutch*) expressing a growth hormone construct exhibited extraordinary growth (Devlin et al., 1994), underwent parr-smolt transformation approximately six months before nontransgenic siblings, and some males matured at just two years of age (Devlin et al., 1995b). However, swimming performance of transgenics was poor (Farrell et al., 1997), perhaps because of a developmental delay or from disruption of locomotor muscles or associated support systems, such as the respiratory, circulatory, or nervous systems. Some growth-enhanced fish exhibited abnormalities of opercular (gill cover) morphology that might disrupt respiration and contribute to poor swimming performance. In competitive feeding trials, Devlin et al. (1999) showed that GH transgenesis increases the ability to compete for food, suggesting that transgenic fish might compete successfully with native fish in the wild. Devlin et al. (2001) noted that the greatest response to expression of the transgene was in Coho hybrids of a wild and domesticated strain; hence, the effects of an introduced growth hormone gene might differ among stocks.

In a study posing implications for introgression of transgenes into wild populations, Devlin et al. (2001) examined the fitness effects of expression of a GH construct in both wild and selectively bred commercial rainbow trout (*O. mykiss*) strains. Transgenic wild-strain rainbow trout retained the slender body morphology of the wild-type strain, but their final size at maturity was much larger than that of their nontransgenic ancestors. Both domestic and wild-strain trout exhibited reduced viability; in the domestic strain, all transgenic individuals died before sexual maturation. The tradeoff of size (and likely mating success) and decreased viability parallels the case modeled by Muir and Howard (1999), and suggests that the viability of a receiving population might be compromised. Devlin et al. (2001) noted that the greatest response to expression of the transgene was in hybrids of a wild and domesticated strain; hence, the effects of an introduced growth hormone gene might differ among stocks. The importance of genetic background on expression of growth hormone was demonstrated also by Siewerdt et al. (2000a,b) and Parks et al. (2000a,b). While indicative that risk issues must be regarded with seriousness, the growing collection of empirical risk assessment studies of transgenic salmonids does not yet provide a body of data useful for parameterizing a model useful for predicting the likelihood that transgenes would become permanently introgressed into wild or feral salmon populations.

However, many of the same physiologic and behavioral differences seen in GE salmon can be induced by using growth hormone implants (Johnsson et al., 1999). As such, implanted fish can model the effects of the transgene and allow the fish to be safely tested in native habitats—an experiment that would be hazardous with GE fish. Working with brown trout (*Salmo trutta*), Johnsson et al. (1999) showed that survival of GH-implanted trout did not differ from that of controls under field conditions with natural predation levels. They concluded that GH-manipulated fish might compete successfully with wild fish despite behavioral differences observed in the laboratory for characteristics such as predator avoidance, foraging ability, and over-winter survival (Johnsson et al., 2000). These results emphasize the need to measure all components of fitness under conditions similar to those found in nature—a task that might not be possible for some species.

Possible environmental hazard pathways posed by the escape of transgenic crustaceans and mollusks into natural ecosystems have not yet been thoroughly considered. Research has not yet assessed ecologic risks posed by production of these organisms. Many freshwater crustaceans, such as crayfishes, are capable of overland dispersal; further, they are produced in extensive systems, where confinement is difficult. Many marine crustaceans have planktonic larvae, thus complicating confinement. Confinement of mollusks can prove difficult at the larval stage (USDA, 1995). Further, because the larval stages drift in the water column before settlement and metamorphosis to the sessile juvenile form, they have great dispersal capability.

The committee's review of ecologic principles and empirical data suggests a considerable risk of ecologic hazards being realized should transgenic fish or shellfish enter natural ecosystems. In particular, greater empirical knowledge is needed to predict the outcome should transgenes become introgressed into natural populations of aquatic organisms.

NEED FOR MORE INFORMATION CONCERNING RISK ASSESSMENT AND RISK MANAGEMENT

Many critical unknowns complicate risk assessment and risk management of genetically engineered animals. Greater knowledge in these areas would support an informed judgment of whether and how to go forward with approval for marketing particular genetically engineered animals. For example, results of well-designed, interdisciplinary studies could prove useful for parameterizing net fitness-based models used for predicting whether transgenic genotypes would persist in natural populations. Should GE animals be approved, postcommercialization monitoring would provide a check on the utility of predictive models, suggest improved means of risk management, and support adaptive management of GE animals (Kapuscinski et al., 1999; Kapuscinski, 2002). More information supporting risk assessment and risk management also would support regulatory decision-making, and it would promote public confidence in the environmental safety of genetically engineered animals.

6

Animal Health and Welfare

INTRODUCTION

The effects of genetic engineering on animal health and welfare are of significant public concern (Mench, 1999). Ideas about animal welfare are shaped by cultural attitudes toward animals (Burghart and Herzog, 1989), and animal welfare has proven difficult to assess because it is so multifaceted and involves ethical judgments (Mason and Mendl, 1993; Fraser, 1999). The committee considered the following animal welfare aspects of transgenic and cloning technologies: their potential to cause pain, distress (both physical and psychologic), behavioral abnormality, physiologic abnormality, and/or health problems; and, conversely, their potential to alleviate or to reduce these problems. Both the effects of the technologies themselves and their likely ramifications were addressed.

REPRODUCTIVE TECHNOLOGIES

Reproductive manipulations, including superovulation, semen collection, artificial insemination (AI), embryo collection, and embryo transfer (ET), are used in the production of both transgenic animals and animals produced by nuclear transfer (NT). Commercial livestock breeders also use many of these manipulations routinely (Chapter 1). However, while these procedures do raise animal welfare concerns (Matthews, 1992; Moore and Mepham, 1995; Seamark, 1993), these generally are not specific to the production of genetically engineered animals. Few of these procedures have received systematic study from the perspective of animal welfare (Van der Lende et al., 2000).

Handling and restraint can be distressful to farm animals (Grandin, 1993) but are essential for almost all husbandry procedures, including those involving reproductive manipulation. Certain reproductive manipulations (e.g., the administration of injections to induce ovulation) can cause additional transient distress, as can electroejaculation.

AI and embryo collection and transfer present a range of animal welfare issues depending on the species used. In cattle, these procedures can be accomplished with minimally invasive non-surgical procedures—the latter under epidural anesthesia. However, in sheep, goats, and pigs these manipulations involve surgical or invasive procedures (laparotomy or laparoscopy), and hence the potential for operative and postoperative pain. In poultry species the hen is killed in order to obtain early-stage embryos. In fish, eggs and milt might be hand-stripped in some species (causing handling discomfort), while in others the males or females must be killed to obtain eggs and/or sperm.

Since breeding livestock are valuable, they might be subjected to these reproductive manipulations repeatedly during their lifetime. In particular, because of the problems involved in screening microinjected embryos prior to implantation to ensure that they actually are carrying the transgene of interest (Eyestone, 1994), recipient cows might be subject to transvaginal amniocentesis for genotyping; nontransgenic fetuses (or male fetuses) are then aborted and the cows reused as recipients (Brink et al., 2000). While this limits the number of recipient animals used, it also raises welfare concerns over the repeated exposure of individual animals to procedures likely to cause pain and distress.

Replacements for, or alternatives to, some reproductive manipulations are available (Moore and Mepham, 1995; Seamark, 1993). For example, a method has been devised for non-surgical embryo transfer in pigs, and ova for some purposes can be obtained from slaughterhouses, which eliminates the need for manipulation of live donor livestock females. The use of nuclear transfer to produce transgenic animals could eliminate the problem of repeated elective abortion and reuse of recipient animals, since cell populations with specific genotypes or phenotypes could be selected before embryo reconstruction (Eyestone and Campbell, 1999).

IN VITRO CULTURE

The development of *in vitro* embryo culture techniques has provided an alternative to *in vivo* culture, but ruminants produced by *in vitro* culture methods, whether or not they are carrying a transgene, tend to have higher birth weights and longer gestations than calves or lambs produced by AI (Walker et al., 1996; Young et al., 1998)—a phenomenon referred to as large-offspring syndrome (LOS). Kruip and den Dass (1997) surveyed researchers worldwide who use *in vitro* reproductive technologies with different breeds of cattle, and also obtained data from a controlled study of Holstein–Friesian calves. The data showed that only 7.4 to 10 percent of calves produced by AI or ET weighed more than 50 kilograms (kg) and only 0.3 to 4.1 percent weighed more than 60 kg, while 31.7 percent of calves produced by *in vitro* procedures (IVP) weighed more than 50 kg and 14.4 percent weighed more than 60 kg. LOS animals have more congenital malformations and higher perinatal mortality rates, although the incidence and severity of the effects reported vary widely among studies (Van Reenen et al., 2001). The range of abnormalities reported includes skeletal malformations (Walker et al., 1996), incomplete development of the vascular system and urogenital tract (Campbell et al., 1996), immune system dysfunction (Renard et al., 1999), and brain lesions (Schmidt et al., 1996). Even when IVP calves are not excessively large, however, they seem to be less viable and more often experience problems like double-muscling, leg and joint problems, hydroallantois, heart failure, enlarged organs, and cerebellar dysplasia (Mayne and McEvoy, 1993; Schmidt et al., 1996; Kruip and den Dass, 1997). In a large-scale study, van Wagtendonk-de Leeuw et al. (1998) found that 3.2 percent of calves born after IVP showed congenital abnormalities as compared to only 0.7 percent of calves produced by AI. Hydroallantois and abnormal limbs and spinal cords were especially prevalent.

The mechanism(s) responsible for these effects are unknown, but chromosomal abnormalities and disturbances in the regulation of early gene expression and in communication between the fetus and the recipient mother have been implicated (Barnes, 1999; Van Reenen et al., 2001). Cows carrying fetuses produced by IVP show abnormal placental development (Bertolini, 2002). Culture conditions are associated with LOS and other developmental abnormalities, and changing culture conditions (e.g., by not using fetal calf serum and not co-culturing with somatic cells) can help to decrease the rates of LOS and perinatal mortality (Sinclair et al., 1999; Van Wagtendonk-de Leeuw et al., 1998). Oocyte quality also might play a role in LOS and other developmental abnormalities (Kruip et al., 2000).

Because of LOS, difficult calvings (dystocia) can be a problem. The mean rate of dystocia across the five breeds represented in the Kruip and den Dass (1997) dataset was 25.2 percent for IVP-produced animals. In the population of Holsteins studied by Kruip and den Dass, dystocia scores were higher (3.05) in IVP than in AI (2.44) or embryo transfer (ET; 2.74) calves, indicating a more difficult delivery in cows carrying IVP fetuses; 14.4 percent of IVP-produced calves died perinatally as compared to 6.6 percent of ET or 6.1 percent of AI calves, and 13 percent of IVP calves were delivered by emergency Cesarean section, as opposed to 0.9 percent of calves produced by standard AI techniques. Because of this, it is becoming more common to deliver IVP offspring by elective caesarian section (Eyestone, 1999). Again, the number of times that this procedure should be performed on any individual animal during her lifetime is an issue of concern. The selection of older, higher parity cows as recipients is important to decrease the incidence of dystocia.

There also is a potential for IVF to have longer-term effects, although detailed data for livestock are lacking (Van Reenen et al., 2001). Even though they are heavier at birth, and might have enlarged organs, IVP-produced bulls seem to have normal semen quality and heifers show normal reproductive maturation (Van Wagtendonk-de Leeuw et al., 2000). IVP calves have normal growth rates and slaughter weights (Farin and Farin, 1995; McEvoy et al., 1998). Studies with mice, however, have shown that *in vitro* manipulation can result in long-term phenotypic changes (Reik et al., 1993), including retarded growth and abnormal DNA methylation patterns; these changes can be transmitted to the offspring (Römer et al., 1997). Intracytoplasmic sperm injection (ICSI) is under development for fertilizing livestock embryos (Chapter 1), and ICSI procedures have been combined with microinjection to produce transgenic animals (Perry et al., 1999). A concern is that, since the normal fertilization method of sperm and egg membrane fusion is bypassed—as is the sperm selection that normally would take place in the female reproductive tract (Galli and Lazarri, 1996)—embryos can be produced from abnormal sperm (Liu et al., 1995), possibly resulting in abnormal offspring.

EFFICIENCY OF PRODUCTION AND NUMBER OF ANIMALS NEEDED

Microinjection (Chapter 2) is an extremely inefficient method for producing transgenic offspring. Although the success of the method varies by species and gene construct, it has been estimated that less than one percent of microinjected livestock embryos result in transgenic offspring, and, of those, typically fewer than half actually express the transgene (Pursel et al., 1989; Rexroad, 1994). Ebert and Schindler (1993) reported efficiencies of between 0 to 4 percent for production of transgenic pigs, cattle, sheep, and goats. About 80 to 90 percent of the mortality occurs very early during development, before the eggs are even mature enough to be transferred to the recipient female (Eyestone, 1994), but postnatal mortality also occurs (Pursel et al., 1989).

Even if an individual does express the transgene, it might not be transmitted to subsequent generations. Approximately 30 percent of transgenic mice are mosaics, which means that they carry the transgene in only some of their cells (Wilkie et al., 1986). High rates of mosaicism are observed in other animals as well (e.g., fish, Hallerman et al., 1990; Gross et al., 1992). In one study involving transgenic cattle, seven out of eight transgenic founder males produced by pronuclear DNA injection were mosaics (Eyestone, 1999). Mosaic founder animals might not pass the transgene to their offspring at all, or they might transmit it at a normal or reduced rate.

In mice and pigs, the inefficiency associated with microinjection can be compensated for to a great extent by implanting recipient females with multiple embryos. In cattle, however, this can result in difficult births as well as masculinization of the female offspring (freemartinism) if both a male and a female embryo are transferred. For this reason, embryos usually are cultured temporarily *in vitro* or in recipient cow, sheep, or rabbit oviducts until the stage at which longer-term viability can be established (Eyestone, 1994). If cows are used, these developed embryos need to be recovered and then transferred to the recipient animals. Although this technique requires the use of additional animals for the "culturing" stage, it can reduce the number of recipient cows needed by up to 90 percent.

MUTATIONS

Because inserted DNA can insert itself into the middle of a functional gene, insertional mutations that alter or prevent the expression of that functional gene inadvertently might be generated (Chapter 2). Meisler (1992) estimates that 5 to 10 percent of established transgenic mice lines produced by microinjection have such mutations, and it is likely that similar rates would be found in microinjected livestock. Most (about 75 percent) of these are lethal prenatally, but those that are not are responsible for an array of defects in mice, including severe muscle weakness, missing kidneys, seizures, behavioral changes, sterility, disruptions of brain structure, neuronal degeneration, inner ear deformities, and limb deformities. Individuals with such mutations can vary enormously with respect to the degree and type of impairment shown. And because many insertional mutations are recessive, their effects do not become obvious until the animals are bred to transgenic relatives (Chapter 2). For example, although mice engineered with a transgene for herpesvirus thymidine kinase were normal, their offspring that were homozygous for the transgene had truncated hind limbs, forelimbs lacking anterior structures and digits, brain defects, congenital facial malformations in the form of clefts, and a greatly shortened life expectancy (McNeish et al., 1988).

Many of the problems associated with random-site integration, including insertional mutagenesis, could be circumvented by gene targeting (Chapter 2), which allows for the controlled integration of transgenes into predetermined loci within the genome. In addition to site-specific transgene insertions, gene targeting also permits the removal (knockout) and replacement of existing genes. However, problems with the expression of inserted genes still can arise, while the phenotypic consequences of knocking out a gene will depend upon the function of that gene.

GENE EXPRESSION

Animal welfare problems also can arise because of poorly controlled expression of the introduced gene (Chapter 2). Many transgenic animals either do not express the inserted gene, or show variable or uncontrolled expression (Seamark, 1993; Eyestone, 1999; Niemann et al., 1999), although the percentage of inappropriate expression might be decreasing as transgenic technologies are refined. It must be noted that earlier experiments with transgenic growth hormone in pigs used metalothionine promoters. Current approaches use more appropriate promoters with greatly reduced abnormalities, although with methods of pronuclear injection, there are still problems and variability.

The most frequently cited example of welfare problems arising from inappropriate transgene expression is that of the so-called Beltsville pigs, which were engineered with a gene for human growth hormone in an attempt to improve growth rate and decrease carcass fat content (Pursel et al., 1987). Backfat thickness was reduced and feed efficiency was improved, although growth rate was not increased. However, the pigs were plagued by a variety of physical problems, including diarrhea, mammary development in males, lethargy, arthritis, lameness, skin and eye problems, loss of libido, and disruption of estrous cycles. Of the 19 pigs expressing the transgene, 17 died within the first year. Two were stillborn and four died as neonates, while the remainder died between two and twelve months of age. The main causes of death were pneumonia, pericarditis, and peptic ulcers. Several pigs died during or immediately after confinement in a restraint device (a metabolism stall), demonstrating an increased susceptibility to stress. Similar problems are seen in mice transgenic for human growth hormone (Berlanga et al., 1993).

Problems due to growth hormone expression also can be seen when the inserted gene comes from the same, or a closely related, species. For example, sheep in which ovine growth hormone inappropriately is expressed are lean but diabetic (Nancarrow et al., 1991; Rexroad, 1994). In salmonids transgenic for fish growth hormone (Devlin et al., 1995a), the largest transgenic fish have growth abnormalities of the head and jaw. Fish with the highest early growth performance are affected the most and have difficulty eating. As a result, growth of these fish is retarded relative to other transgenics at 15 months of age, and they die prior to maturation. Thus, the severity of morphologic abnormalities is correlated with initial growth rate, although not all transgenic fish display abnormalities. Devlin et al. (1995b) also observed that transgenic coho salmon exhibit cranial deformities and opercular overgrowth. After one year of development, the overgrowth of cartilage in the cranial and opercular regions of the fish with this atypical phenotype becomes progressively more severe and reduces viability. Further, all F_1 progeny were deformed seriously, with excessive cartilage growth in the cranium, operculum, and lower jaw, and they had low viability. The deformities in the offspring were more severe than those observed in their parents at the same age. Devlin attributed this to the mosaicism between founder and F_1 generation, with elevated levels produced in the F_1. Devlin et al. (1995a) concluded that the best optimal long-term stimulation is achieved in transgenic individuals that show intermediate levels of initial growth enhancement.

As in mice, the genetic background of particular selected strains of farm animals probably is important in determining the severity of the defects associated with the transgene. Pursel et al. (1989) speculated that the deformities found in the Beltsville pigs would have been less severe if the foundation stock had been selected for leg soundness and adaptation to commercial rearing conditions.

UNIQUENESS OF TRANSGENIC ANIMALS

Because there can be so much variation in the sites of gene insertion, the numbers of gene copies transferred, and the level of gene expression (Chapter 2), every transgenic animal produced by microinjection is (theoretically, at least) unique in terms of its phenotype. Pigs transgenic for growth hormone, for example, vary enormously in the number of DNA copies that they have per cell (from 1 to 490) and in the amount of growth hormone that they secrete (from 3 to 949 nanograms per milliliter, or ng/ml). Half of pigs transgenic for a gene (c-ski) intended to enhance muscle development experienced muscle weakness in their front legs, and in general the degree and site of muscle abnormality in these pigs varied considerably from one individual to another (Pursel et al., 1992).

This variability makes the task of evaluating the welfare of transgenic animals particularly difficult, since adverse effects almost are impossible to predict in advance, and each individual animal must be assessed for such effects. Van Reenen and Blokhuis (1993) describe the difficulties involved in such assessments. In most cases, deleterious phenotypic changes in transgenic farm animals—particularly animals transgenic for growth hormone or other growth promoting factors—have been easy to detect because they cause such gross pathologies. However, more subtle effects also are possible. Growth hormone, for example, has many systemic effects, including effects on the efficiency of nutrient absorption, fecundity, and sexual maturation (Bird et al., 1994). Growth hormone constructs in salmonids have been shown to influence smoltification (Saunders et al., 1998), gill irrigation, disease resistance, body morphometry (Devlin et al., 1995a,b), pituitary gland structure (Mori and Devlin, 1999), life span (Devlin et al., 1995a,b), and larval developmental rate (Devlin et al., 1995b).

Gene insertion and removal also can have effects on behavior—sometimes subtle. For example, growth hormone constructs in fish have been found to affect swimming ability (Farrell et al., 1997), feeding rates (Abrahams and Sutterlin, 1999; Devlin et al., 1999), and risk-avoidance behavior (Abrahams and Sutterlin, 1999). Some types of knockout mice also have been found to exhibit behavioral problems, such as increased aggressiveness and impaired maternal and spatial behaviors (Nelson, 1997) that are not immediately apparent, but that significantly could affect housing and care requirements.

Sometimes adverse effects are seen only when animals are challenged in some way. The abnormal stress response of the Beltsville pigs, when restrained, is an obvious example. In addition, some problems might not become evident until later in development. Mice transgenic for an immune system regulatory factor, interleukin 4, develop osteoporosis, but not until about two months of age (Lewis et al., 1993). This emphasizes the importance of monitoring the welfare of founder transgenic animals, and sometimes successive generations, throughout their lifetime using multiple criteria, including behavioral abnormality, health, and physiologic normality (Van Reenen et al., 2001). There has been only a limited number of studies of the welfare of transgenic farm animals to date, and detailed behavioral studies are particularly lacking.

NUCLEAR TRANSFER

Somatic cell nuclear transfer (NT) is a relatively new process (Chapter 1), and currently is very inefficient. High prenatal mortality and developmental abnormality, LOS, perinatal mortality, and abnormal placentation commonly are reported in cloned cattle and sheep (e.g., Wilson et al., 1995; Garry et al., 1996; Wells et al., 1997; Kato et al., 1998; Hill et al., 1999; 2000; De Sousa et al., 2001). Most mortality in cloned offspring appears to occur within the first few days after birth, although later mortality also is seen. Health and welfare problems reported in the immediate postnatal period include respiratory distress, lethargy, lack of a suckling reflex, cardiomyopathy, pulmonary hypertension, hydroallantois, hypoglycemia, hyperinsulinemia, urogenital tract abnormalities, pneumonia, and metabolic problems. However, such problems are not seen universally in cloned animals; many apparently healthy adult cattle, sheep, and goats have been cloned from adult, fetal, and embryonic cells (Lanza et al., 2001; Cibelli et al., 2002). For example, Wells et al. (1999) succeeded in producing 10 healthy calves from 100 transferred NT blastocysts; the calves were not exceptionally large, all had a strong suckling reflex, and only one required veterinary intervention. Lanza et al. (2001) report that the 24 dairy cows surviving from an original group of 30 cloned cattle are in normal physical condition for their stage of production, exhibited puberty at the expected age, have high conception rates after artificial insemination, and show no clinical or immunologic abnormalities.

It is difficult to determine which problems are due to cloning (nuclear transfer) per se, to embryo culture or transfer methods, or to some combination of cloning and culture/transfer methods (Wilson et al., 1995; Kruip and den Dass, 1997; Van Wagtendonk-de Leeuw et al., 1998). There is considerable variation among studies in rates of early embryonic death, perinatal mortality, LOS, and dystocia (Kruip and den Dass, 1997; Cibelli et al., 2002). The incidence of these problems actually is sometimes lower in animals produced by NT than is typical for animals produced by IVP. Varying levels of expertise and proficiency with the relevant techniques certainly could be contributing factors. Because of their economic value, cloned animals would be expected to receive a high level of veterinary oversight and intervention, which could contribute to the higher postnatal survival of cloned animals in some studies. In cases where there are neonatal problems, they might resolve within a few days of birth (Garry et al., 1996).

One possible contributing factor to the high prenatal and neonatal mortality seen in cloned animals is improper epigenetic reprogramming (Young and Fairburn, 2000; Rideout et al., 2001). Cloned animals have abnormal methylation patterns, although the significance of this for embryo development and survival in livestock is unclear. The longer-term effects of cloning and/or improper epigenetic reprogramming on animal welfare have yet to be thoroughly evaluated; as the number of surviving cloned livestock increases, such assessments will be possible. There still is a need for detailed behavioral studies of cloned livestock, since cloning has been shown to result in the impairment of mice in learning and motor tasks, although this impairment is transient (Tamashiro et al., 2000).

Clones produced by fusion of nuclear donor cells with unfertilized eggs are not identical twins, but "genetic chimeras," since almost all cloned livestock studied to date have mtDNA from the recipient egg but not from the donor cell (Evans et al., 1999; Takeda et al., 1999). Whether or not there are potential adverse effects on health and welfare due to having nuclear DNA from one source and mtDNA from another are unknown, although mitochondria are responsible for important cellular functions and mitchondrial type theoretically could affect relevant production traits as well. Of course, each time normal fertilization occurs, nuclear genes from the sperm are introduced into a different genetic mitochondrial environment than existed in the cells of the male providing the sperm, so the mixing of nuclear and mitochondrial genes is ubiquitous in nature.

During normal aging, telomere lengths shorten, and this phenomenon has been associated with cell senescence (Chapter 2). Normal reproductive processes restore telomere lengths in newborns, but there has been concern about whether this same restoration would be seen in animals cloned from adult cells, or whether such animals instead will age prematurely and possibly develop health problems usually seen in older animals. While shortened telomere lengths were seen in one sheep ("Dolly") cloned from adult somatic cells (Shiels et al., 1999), telomere lengths apparently are normal in cattle cloned from adult cells (Lanza et al., 2001; Betts et al., 2001).

BIOMEDICAL APPLICATIONS

In contrast to genetic manipulation of farm animals for production traits, transgenic manipulation for the production of human pharmaceuticals or transplant organs generally is not intended to cause changes that have physiologic effects on the animals themselves. Thus, although unexpected and undesirable phenotypic effects still can occur as a result of gene insertion or cloning technology, there generally are fewer potential animal welfare concerns associated with the production of transgenic farm animals for biomedical purposes than for agricultural purposes (Van Reenen and Blokhuis, 1993).

Pharmaceuticals

Although there is a potential for producing pharmaceuticals in the eggs, blood, urine, or sperm of farm animals (Lubon, 1998; Sharma et al., 1994), the most common method is to produce transgenic cattle or goats that express the protein of interest in mammary tissue. The recombinant protein then is secreted in milk when the female lactates. This poses problems mainly when those proteins either are expressed in non-mammary tissues (so-called ectopic expression) or when they "leak" out of the mammary gland into the circulation (e.g., Lubon, 1998; Niemann et al., 1999). If the protein is active biologically in the species in which it is produced, it can cause pathologies and other severe systemic effects (e.g., Massoud et al., 1996). Rigorous regulation of the expression of the transgene thus is necessary to ensure that the animal welfare consequences of milk-borne pharmaceutical production are minimized, but such regulation currently is difficult to achieve. However, even when a pharmaceutical is confined to the mammary tissue, the expression of particular proteins has been associated with premature lactational shutdown in goats (Ebert and Schindler, 1993) and pigs (Shamay et al., 1992). In pigs, there was evidence that the mammary tissue developed abnormally due to premature expression of the transgene, and that the condition of the mammary gland might have caused lactation to be painful. Similar concerns arise in the case of blood-borne proteins and nutraceuticals (see below) if the products are produced at levels higher than the animal's normal physiologic levels.

Xenotransplantation

In an attempt to prevent hyperacute rejection of pig organs by humans (Chapter 2), pigs have been made transgenic for the expression of human complement proteins, which are involved in regulation of the immune response (Cozzi and White, 1995; Tu et al., 1999; Cozzi et al., 1997; Byrne et al., 1997; Cowan et al., 2000). No phenotypic abnormalities have been reported in pigs as a result of the expression of transgenes for these human proteins, although, since the pigs are produced by microinjection, there are the usual inefficiencies in terms of the number of embryos microinjected relative to the number of transgenic animals born (Tu et al., 1999; Niemann and Kues, 2000).

Research is underway to produce pigs that, in addition to carrying complement transgenes, have both copies of the gene encoding the enzyme that produces the antigen associated with rejection knocked out. The animal welfare implications of this genetic manipulation are unknown; however, the knockout, which causes changes in cellular carbohydrate structure, potentially could have deleterious physiologic effects on the animals (Dove, 2000) and also render them susceptible to infection with human viruses (Chapter 2).

An important animal welfare concern related to xenotransplantation is the management and housing of pigs intended for use as organ sources. To minimize the potential for transmission of disease to human recipients, only specific pathogen free (SPF) pigs are used. SPF research animals are used in other contexts besides xenotransplantation, but their use raises several animal welfare issues. SPF pigs are born by hysterotomy or hysterectomy, and then are reared in isolators for 14 days before being placed in the source herd or in the xenotransplantation facility. The natural weaning age for pigs is about eight weeks (three to four weeks in commercial practice), and piglets subjected to extremely early weaning like this are known to develop abnormal behaviors (Weary et al., 1999). Older pigs intended for testing or organ donation might be housed in social isolation in unusually barren (i.e., easily sanitizable) environments. Pigs are extremely social animals that, when given the opportunity, will spend considerable time each day foraging, and that develop abnormal behaviors in confinement if not given the opportunity to root or build nests. In the United Kingdom, the Home Office Code of Practice (Her Majesty's Government, 2000) for organ-source pigs, while recognizing the importance of maintaining biosecure facilities, nevertheless recommends that such pigs be housed in stable social groups, and provided with environmental enrichment such as straw or other material suitable for manipulation. The Code requires justification if the animals' behavioral needs are to be compromised for a xenotransplantation protocol. There are no comparable standards for pigs intended for xenotransplantation in the U.S., and the lack of standardization of housing and care among U.S. facilities for these pigs is a source of concern. Although there are many forms of environmental enrichment available that are suitable for laboratory-housed pigs (Mench et al., 1998), appropriate methods for organ-source pigs require development and evaluation (Orlans, 2000).

Other Biomedical Applications

Farm animals might be genetically engineered for human biomedical applications other than xenotransplantation or the production of pharmaceuticals. Research is underway, for example, to produce a porcine model of cystic fibrosis, and there already are farm animal models for retinal degeneration (Petters et al., 1997) and neurodegenerative disease (Theuring et al., 1997). As genetic engineering techniques for farm animals improve—particularly such that single base coding changes that are typical of many human genetic diseases can be introduced, and the production and use of farm animal models becomes more economically feasible—it is likely that more models for disease research and toxicity testing will be developed. Discussion of the potential issues raised by these biomedical uses of farm animals is outside the scope of this report. However, the welfare implications will depend upon specific features of the model under study, including any unalleviated pain and suffering associated with the disease process itself, as well as the need for specialized husbandry and veterinary care requirements (Dennis, 2002).

FARMING

If genetic technology becomes more efficient and affordable, the primary farming applications of transgenesis and cloning likely will be to produce animals with increased growth, improved feed conversion, leaner meat, increased muscle mass, improved wool quality, improved disease resistance, and increased reproductive potential. The technology also can be used to produce food of improved nutritional quality (nutraceuticals) or appeal.

The primary difference between traditional breeding and genetic engineering is the speed at which change typically occurs (although naturally occurring mutations and recombination events also can cause rapid and dramatic change), and the single-gene nature of genetically engineered change. Traditional methods of selection are more likely to be subject to the checks and balances imposed by natural selection. Many related and apparently unrelated traits are correlated genetically; thus, selective breeding involves selecting for a whole phenotype rather than a single gene product. Because most production and behavioral traits in livestock are polygenic and our understanding of livestock genomes is poor, few traits can reliably and predictably be engineered or introduced by manipulating only one gene (Moore and Mepham, 1995). For this reason, the production of a line of transgenics will require generations of selective breeding after the introduction of gene constructs into the founder generation to ensure that animals display the desired phenotype with few or no undesirable side effects.

However, it is clear that serious welfare problems also have resulted from traditional breeding techniques. Broiler chickens are a case in point. Breeding for increased growth has led to serious physical disabilities, including skeletal and cardiovascular weakness. A large percentage of broilers have gait abnormalities (Kestin et al., 1992), and these might be painful, making it difficult for the birds to walk to feeders and waterers. In addition, broiler hens must be severely feed restricted to prevent obesity, and this feed restriction is associated with extreme hunger and a variety of behavioral problems, including problems with mating behavior and hyperaggressiveness (Mench, 2002; Kjaer and Mench, in press). Traditional selection of pigs for increased leanness has led to increased excitability during handling (Grandin and Deesing, 1998), and selection for high reproductive rates (either by shortening the interval between births or increasing the number of offspring born) or increased lactation (Chapter 1) also has led to welfare problems. In their report, *The Use of Genetically Modified Animals*, the Royal Society (2001) concluded: "Although genetic modification is capable of generating welfare problems...no qualitative distinction can be made between genetic modification using modern genetic modification technology and modification produced by artificial selection." Several ethical frameworks for evaluating the animal welfare implications of biotechnologies applied to animals have been proposed in an attempt to resolve this difficulty. For example, Rollin (1995) has proposed the use of the "principle of conservation", which states that transgenic and cloned animals developed for agricultural uses should not be worse off than the founder animals or other livestock of the same species under similar housing and husbandry practices.

POTENTIAL ANIMAL WELFARE BENEFITS

Genetic engineering certainly has the potential to improve the welfare of farm animals. Decreasing mortality and morbidity by increasing resistance to diseases or parasites, or decreasing responses to ingestion of toxic plants, are obvious examples of welfare benefits, and an area in which some transgenic research is focused (Müller and Brem, 1994; Dodgson et al., 1999). It also has been pointed out that transgenic animals might receive a higher standard of care than nontransgenic animals because of their greater economic value (Morton et al., 1993). Cloning could be used as a strategy for breed preservation to maintain genes that are important for adaptation and resistance to disease, but equally could result in a further narrowing of the gene pool, with possibly deleterious effects on animal health (Chapter 2).

Improving disease resistance to decrease pain and suffering is an application of transgenic technology that has clear animal welfare benefits. But it should be stressed that animal welfare is multifaceted, and this needs to be taken into account when assessing welfare impacts of the application of any technology—not just biotechnology. Important elements of animal welfare include freedom from disease, pain, or distress; physiologic normality; and the opportunity to perform normal behaviors (Broom, 1993). While reducing disease clearly is beneficial, if this also permits animals to be confined more closely, and thus decreases the opportunity for them to perform their normal behaviors, then the net effect on welfare could be negative.

Genetic engineering also could be used to deal with non-disease related welfare problems. It might be possible, for example, to engineer hens that produce only female offspring (Banner, 1995). This would eliminate the problems associated with surplus male chicks, which are killed at the hatchery. The need for the so-called standard agricultural practices like castration and dehorning also could be reduced or eliminated by genetic engineering. Pigs are castrated to prevent boar taint in the meat, but this trait is strongly linked (genetically) and thus is amenable to genetic manipulation. Similarly, horns on cattle, which are removed because they cause injuries to humans and other cattle, are the result of a single gene that could be knocked out by genetic manipulation without affecting other desirable performance traits; genetically polled (hornless) breeds of cattle already are available, and are produced by selective breeding.

COSTS VERSUS BENEFITS

In making assessments about the production of genetically engineered animals for farming, costs and benefits need to be weighed carefully. When expression of growth hormone is regulated appropriately in transgenic pigs, for example, the increases shown in growth and feed efficiency are modest, and are similar to the increases that can be attained simply by injecting pigs with porcine growth hormone (Pursel et al., 1989; Nottle et al., 1999). Pursel et al. (1989) suggest that centuries of selection for growth and body composition might limit the ability of the pig to respond to additional growth hormone. Indeed, it is possible that we already have pushed some farm animals to the limits of productivity that are possible by using selective breeding, and that further increases only will exacerbate the welfare problems that have arisen during selection.

The potential for reduction in genetic diversity in agricultural species also is posed by inappropriate application of certain biotechnologies (Chapter 1). Transgenesis raises such concerns because each transgene integration event results in a genetically unique potential founder and only one founder normally is used to found a transgenic line. This can result in a profound genetic bottleneck unless genetic variability is restored to a production line by purposeful utilization of a mating strategy involving backcrossing of the transgenic line to a large number of distinct, presumably nontransgenic, mates. The effects of cloning are more difficult to anticipate because competing processes are at issue. On the one hand, cloning by its nature produces identical copies of a particular individual, reducing genetic variability relative to what would have been transmitted via conventional breeding. On the other hand, cloning makes it possible to save and utilize genetic variability that would not otherwise be available. For example, cloning could be employed to utilize the genetic resources from a steer that had proven to be a high performing individual. Cryopreserved cells could be utilized as donor material. Moreover, cloning is a tool that actually can be used to increase/maintain genetic variance in some situations quite independently of exploiting castrates (Seidel, Jr., 2001). The tradeoff between the competing processes of loss and gain of genetic variance would be case-specific, and it is hard to quantify in the absence of simulation modeling with validation from field observations. Whatever the mechanism causing it, loss of genetic diversity could limit the potential for future genetic improvement of breeds by selective breeding or biotechnologic approaches. Furthermore, disease could spread through susceptible populations more rapidly than through more genetically diverse populations.

A particularly serious concern that arises is susceptibility of species with low genetic diversity to infectious disease. Diversity of animal populations—particularly at major histocompatibility (MHC) loci—is a major factor preventing spread of disease (particularly viral disease; Xu et al., 1993; Schook et al., 1996; Kaufman and Lamont, 1996; Lewin et al., 1999). Different MHC types recognize different viral or bacterial epitopes encoded by pathogens for presentation to the immune system. In genetically diverse populations, pathogens can evade the immune response only if they adapt to each individual MHC type following transmission from one individual to another. The requirement for this evolutionary process provides a population of animals with significant protection against the spread of infection. Pathogens can evade host immune response more easily in genetically uniform populations (Yuhki and O'Brien, 1990). The consequences of the failure of immunorecognition are illustrated by the deadly epidemics of diseases—such as measles—spread by initial contact between Europeans and isolated New World populations that lacked adequate MHC diversity. Not only could enhanced susceptibility create significant risk for spread of "new" infectious diseases in "monocultures" of cloned or highly inbred animal populations, it also could create new reservoirs for the spread of zoonotic infections—like new strains of influenza—to humans.

Concerns Related to Scientific Uncertainty, Policy Context, Institutional Capacity, and Social Implications

Most of the concerns about animal biotechnology addressed in this report involve potential impacts on human or animal health and the environment. These are among the specific science-based concerns that regulatory agencies might consider in formulating regulatory policies and in making decisions about specific applications of biotechnology to animals. To address these concerns in a scientifically sound and publicly acceptable manner, however, it also is important to consider the scientific, policy, institutional, and social context in which the concerns about animal biotechnology are arising and will be addressed. This chapter does not attempt to address these issues exhaustively, but enough to convey the broader intellectual, public policy, and social dimensions of how society likely will respond to the scientific concerns raised in this report, and to underscore the need for public participation in decisions about animal biotechnology. Nonscientific concerns should not alter scientific analysis, but they will inevitably and properly influence the procedural framework within which scientists address questions that have regulatory consequences, and they will shape the public policy response to science-based health and environmental concerns.

SCIENTIFIC UNCERTAINTY

Scientific uncertainty is an important part of the context for animal biotechnology. Uncertainty is a common feature of regulatory decision-making.

Indeed, the essence of regulatory decision-making on health and environmental issues is to make judgments, in the face of uncertainty, about whether established standards have been met. Although it is impossible to prove the safety of a product or technologic application with complete certainty, regulatory scientists (scientists who are responsible for scientific evaluations for a regulatory agency) usually operate within established protocols for evaluating the safety of products or technologies and manage uncertainty by applying safety factors when estimating risks and by identifying additional studies that can provide data to reduce uncertainties. In the case of at least some applications of biotechnology to animals, however, scientific uncertainty will be a particular concern, due to the novelty of the health and environmental questions posed, and due to the lack of established scientific methods for answering them.

Uncertainties can be placed in three categories—statistical, model, and fundamental. These categories generally correspond to technical, methodologic, and epistemologic considerations, respectively, which also can be described as inexactness, unreliability, and insufficient knowledge (Funtowicz and Ravetz, 1992).

Statistical uncertainty—usually centered around the value of a single variable—is reduced most easily by additional data collection, leaving residual uncertainty that can be quantified. For example, the impact of bovine somatotropin (BST) use on milk production, IGF-1 levels in milk, or the incidence of mastitis in treated animals can be studied rather easily, and the probability distribution of values for each of these variables can be determined.

Model uncertainty results from not fully understanding interactions among variables in models used to predict the behavior of multivariate systems when one or more variables are changed. Model uncertainty inherently is more difficult to reduce and to quantify than statistical uncertainty. For example, the potential of transgenic fish to enter the natural environment and alter the marine ecology is a new concern for regulators and scientists that brings into play multiple variables and interactions; this issue poses novel scientific questions, and requires new data collection protocols and methods of analysis. Similarly, a transgene might have pleiotropic effects on multiple fitness traits, making the net effect difficult to predict. In Japanese rice fish engineered with a growth hormone transgene, for example, the disadvantage of a reduction in juvenile viability might be more than offset by the advantages of earlier sexual maturity and an increase in female fecundity relative to wild type (Muir and Howard, 2001). The Trojan gene example in Chapter 5 also shows how, as model uncertainty increases, an even more fundamental kind of uncertainty begins to appear.

Fundamental uncertainty results from indeterminacy, ignorance, or ignorance-of-ignorance. In the case of novel technologies, existing models might not apply. Moreover, if we are ignorant of the potential existence of a particular hazard, we might fail to consider it at all when attempting to estimate the potential harms or benefits of an activity. Molecular breeding by DNA shuffling, for example, will result in at least some outcomes that fundamentally are uncertain and always will be virtually impossible to predict. We will remain ignorant of them until they occur, and even then, might only identify them if we search in sensitive ways. Attempts to estimate the probability of harm (or benefit) from such a fundamentally uncertain activity must be undertaken with great care since ignorance-of-ignorance might lead to serious errors.

The kind and degree of scientific uncertainty have implications for the processes agencies use or devise to reach sound and publicly acceptable decisions (see Box 7.1). In the case of model uncertainty, for example, more effort might be required to engage a broad scientific community in consensus building about protocols for evaluating hazards and to air specific risk assessments publicly. Scientific uncertainty—especially in the model and fundamental categories—also might have implications for public and private research priorities, market-entry standards and processes, and other regulatory policies such as the need for post-approval monitoring and research.

BOX 7.1
Error Bias

Error biases are important determinants of conclusions drawn from the interpretation of scientific data and, therefore, often have a direct influence on public policy. The impact of error biases might be particularly important when the analysis of complex systems requires numerous assumptions and simplified models in order to attempt to predict system behavior. For new technologies, which might be characterized by fundamental uncertainties, we might be ignorant of what to look for or how to frame a research question, setting the stage for surprises and unpredicted impacts. Error biases usually are set by agreed-upon convention, and hypothesis testing commonly favors Type II errors (false negatives) over Type I errors (false positives). That is, a null hypothesis commonly asserts that there is no relationship between an action and a response in a system, and highly significant evidence typically is required as a basis for rejecting the null. In addition, asking the wrong question or failing to ask the right question, sometimes called a Type III error, also is problematic when dealing with novel technologies. The failure to identify a hazard when it exists (Type II error) might lead to policies that are not protective of health or the environment. Conversely, identifying a technology or product as hazardous when it is not (Type I error) might lead to unnecessary, burdensome regulation or the failure to adopt something useful. With respect to emerging animal biotechnologies, the committee acknowledges that, for many applications, hazard/safety data are sparse, and, in many studies, the number of individuals, populations, or models examined is small. Uncommon or less common events are less likely to rise to statistical significance and might not even be identified in such a limited dataset, resulting in a bias toward Type II error in data interpretation if these limitations are not kept in mind. The likelihood of a Type III error (asking the wrong question) will depend entirely upon how comprehensively and systematically examiners look for the potential impacts of the various technologies.

POLICY CONTEXT

In addition to posing new scientific questions and increasing scientific uncertainty, the novelty of some of the concerns posed by animal biotechnology raises a policy question about the meaning of the health and environmental safety standards under which the scientific questions will be addressed. The U.S. Food and Drug Administration (FDA) has said that it intends to regulate transgenic fish and other transgenic animals under the new animal drug provisions of the Federal Food, Drug, and Cosmetic Act. This law directs the FDA to license animal drugs that the sponsor has demonstrated to be safe for human and animal health and effective for their intended use, which typically is therapeutic or to promote animal growth and productivity. The meaning of these standards is well understood in the context of conventional animal drugs, and there are well-established scientific protocols for collecting relevant data and evaluating whether the standards have been met. The safety issues are relatively straightforward because they focus on the health of the treated animal and the safety of edible tissues derived from the animal.

It might be less clear what these safety standards mean in the context of animal biotechnology. The FDA has said, for example, that it considers the animal safety aspect of the animal drug standard to apply not only to the transgenic animal, but also to wild fish and other animals in the environment that might be affected by the release of the transgenic animal. On this basis, the FDA says it will regulate the environmental impacts of transgenic fish, such as the transgenic salmon currently under FDA review (OSTP, 2001). What does "safe" mean in this context? What environmental impacts and direct or indirect impacts on the health of wild fish and fish populations fall within the scope of the statutory safety standard for animal drugs? What degree of precaution is appropriate in evaluating these impacts? How will the novel scientific uncertainties associated with environmental hazards posed by transgenic fish be managed in regulatory decisions under the statutory safety standards for animal drugs? How will the expertise and perspectives of scientists and other stakeholders be considered by FDA under the animal drug licensing process, which is closed to public participation?

It is beyond the scope of this study to address or attempt to answer these policy questions. They are relevant, however, to identifying the scientific concerns over animal biotechnology that government scientific reviewers and regulators will have to address, and to determining the scientific approaches that will be adequate to address them. These policy questions are relevant because the meaning and scope of the safety standard are prime determinants of what agencies must consider a relevant scientific concern. Moreover, the quantity and quality of the scientific data required to address an identified safety concern, as well as whether available scientific protocols are adequate to collect the needed data, are a direct function of the degree of precaution the regulatory agency considers appropriate and the degree of scientific uncertainty it deems acceptable. These are questions yet to be addressed and resolved in the context of transgenic animals.

Another policy-related concern is the regulatory environment with respect to animal welfare. The animal welfare regulatory system in the United States is complex. Livestock used for biomedical research are covered under the Animal Welfare Act Regulations (AWRs) and the Public Health Service (PHS) Policy, which also covers research projects funded by national research institutes like the National Institutes of Health. Fish and birds used in biomedical research funded by national research institutes also are covered under the PHS policy, but are not regulated by the U.S. Department of Agriculture (although the USDA has announced its intention to regulate birds). Both the PHS policy and the AWRs require that animal research protocols be reviewed and approved by an Institutional Animal Care and Use Committee prior to their initiation. The intent of such review is to ensure that animal pain and distress are minimized, that alternatives have been investigated, and that the minimum number of animals necessary to achieve research goals is used. There is no such requirement for review of production-related (i.e., food and fiber) research protocols involving animals, although voluntary standards for such review are available (in the so-called *Ag Guide*; FASS, 1999). This two-tier system means that research projects involving biomedical uses of genetically manipulated farm animals for xenotransplantation and pharmaceutical production will be reviewed for their potential impacts on animal welfare, and the animals involved in those projects will be subject to some type of oversight. Those projects directed toward genetic manipulation for improved food or fiber production, on the other hand, might or might not be subject to such review and oversight, depending upon whether or not the institution at which the research is conducted has chosen to adopt the *Ag Guide* or a similar set of standards.

An additional concern relates to the effect of the patent process on animal welfare. If technologies to reduce the number of animals used in transgenesis, or to reduce the incidence of developmental abnormalities, become available but are patented, those technologies might not readily be accessible to producers and marketers of genetically engineered animals. Less sophisticated technologies that have more negative impacts on animal welfare might thus continue to be used for the production of transgenic animals.

INSTITUTIONAL CAPACITY

The institutional framework for regulation of animal biotechnology affects how science-based concerns about the technology will be identified and resolved. The committee has identified features of the institutional framework that raise concerns, including the multiplicity of agencies and statutes potentially involved in regulatory oversight of animal biotechnology and the legal and technical capacities of the agencies to address some of the novel questions posed by the technology.

Agencies and Statutes

Appendix B lists the many components of the federal government that might have jurisdiction over some aspect of animal biotechnology. They include potentially four different centers within the FDA and two agencies in the U.S. Department of Agriculture that have some jurisdiction over the animal and/or human health impacts of animal biotechnology, depending on the nature and intended use of the product involved. In addition, some of these components of government, such as the FDA's Center for Veterinary Medicine (CVM), have said that they will regulate the environmental impacts of the technology, but there are additional agencies that also might have a role on environmental issues, such as the U.S. Army Corps of Engineers (ACE), the Environmental Protection Agency (EPA), the Fish and Wildlife Service (FWS) in the Department of the Interior, the National Marine Fisheries Service (NMFS) in the Department of Commerce, and various state-level environmental and natural resource agencies.

Each of these agencies operates under its own distinct statutory mandate and mission, which necessarily influences the nature of the scientific questions that they will consider important in carrying out their responsibilities. In the case of transgenic fish, for example, the CVM claims primary jurisdiction over environmental issues, but the ACE has jurisdiction under the Rivers and Harbors Act over the siting of aquaculture facilities in navigable waters, where net pen salmon facilities commonly are found. Under this act, which gives the ACE broad discretion on whether and how to act on environmental matters, the ACE balances a host of concerns, including conservation and environmental impacts, and, like CVM, is subject to its own assessment requirement under the National Environmental Policy Act (NEPA) in making siting decisions. The FWS and NMFS have regulatory roles under the Endangered Species Act to the extent that the siting of an aquaculture facility or any other government action could affect an animal on the endangered species list, such as Atlantic salmon. And the EPA already has invoked its Clean Water Act authority to regulate discharges from salmon aquaculture facilities in Maine (Lubber, 2000), and potentially could do so again with transgenic fish facilities.

Multiple agencies also are potentially involved in food safety aspects of animal biotechnology. While CVM claims jurisdiction over the genetic transformation of livestock under its animal drug authority, meat from slaughtered animals will be inspected by the Food Safety and Inspection Service of the U.S. Department of Agriculture. At the federal level, milk is under the jurisdiction of a different component of FDA, the Center for Food Safety and

Applied Nutrition, which has the FDA's core expertise in food safety and nutrition. Milk inspection, however, is handled primarily at the state level.

The multiplicity of agencies and statutes potentially involved in regulating the safety and environmental aspects of animal biotechnology is a concern for scientists and other stakeholders, who will be seeking clarity about the scientific standards, data requirements, and analytical approaches to be applied in making market entry decisions. Without this clarity, it will not be possible to gather the necessary data with efficiency, and with confidence that the data will be scientifically sufficient and meet the government's regulatory needs. Moreover, without clarity concerning scientific requirements and the allocation of responsibilities among the federal agencies, the public will have difficulty understanding, evaluating, and ultimately, gaining confidence in the government's decisions.

The committee notes a particular concern about the lack of an established regulatory framework for the oversight of scientific research and commercial application of biotechnology to arthropods. As discussed in Chapter 5, genetically engineered insects could pose substantial and difficult-to-assess environmental hazards, and could present especially difficult containment issues, yet research and commercial experimentation is proceeding without any regulatory oversight (Hoy, 2001).

In addition to the potential lack of clarity about regulatory responsibilities and data collection requirements, the committee notes a concern over the legal and technical capacity of agencies to address potential hazards, particularly in the environmental area. The CVM's statute—the animal drug provisions of the Federal Food, Drug, and Cosmetic Act (FDC Act)—for example, was enacted to address the safety of animal drugs with respect to the treated animals and any residues that remain in edible tissue, such as meat, milk, and eggs. The statute seems well designed for this purpose, and the CVM has extensive experience and expertise in addressing these safety issues. The FDC Act is not, however, an environmental statute. It thus is unclear whether the "health of man or animal" language in the FDC Act's definition of the safety standard for animal drugs will be broad enough to sustain FDA's regulatory authority over broad, systemic effects of animal biotechnology on ecosystems, such as harms to centers of origin and other genetic resources, or a decline in the resilience of a fish community (Kapuscinski, 2000). Nor is the CVM an environmental agency by mandate or tradition. Moreover, the agency lacks expertise in specialized areas that are relevant to assessing the environmental impacts of animal biotechnology, such as marine ecology and evolutionary biology.

The committee's concern in this area is underscored by the novelty of the environmental impact questions potentially posed by animal biotechnology and the methodologic uncertainties about how to assess and manage those impacts. Assessing the environmental and ecologic risk posed by a transgene release is complex in part because multiple outcomes are possible for any transgene. This is to some extent inherent in the nature of random insertion of DNA. Each gene construct used to transform each species, or even the same construct in different fish of the same species, might produce a unique risk of gene spread (Chen et al., 1994). Several reasons underlie such variable outcomes, including alternative insertion sites and copy number of the transgene, genetic regulatory mechanisms, the effect of transgenes on the target trait as well as effects on other traits, and the scale and frequency of their introduction into the natural population (Kapuscinski and Hallerman, 1990; 1991). Thus, it is necessary to consider whether, because of random gene insertion, each founder poses a unique risk.

The complexity of predicting environmental impacts is compounded by the nearly infinite number of direct and indirect biotic interactions affecting gene spread that occur in nature, and the fact that populations of a species can evolve in response to a hazard. Predictive, fitness-related models have been developed (Muir and Howard, 2001; 2002a; Hedrick, 2001), but they have not been tested in a regulatory context, and they involve scientific issues different from those normally addressed by the CVM.

Commercial application of animal biotechnology might require adoption of containment strategies to reduce the risk of gene spread and adverse environmental impact. In the case of transgenic fish, mechanical (e.g., screens at water inlets and outlets), physical (e.g., lethal pH or temperatures applied to rearing unit effluent water), and biologic containment approaches have been developed and might be applicable to minimize unintentional release into the environment (Devlin and Donaldson, 1992). Biologic containment, which might be especially important due to the high likelihood of escape from mechanical or physical containment, can be achieved through various means, including sterility by induced triploidy (Benfey, 1999) and by a mix of hormonal and transgenic methods (MacLean and Penman, 1990; Devlin and Donaldson, 1992). There remain, however, uncertainties about the efficacy of various containment measures and what degree of efficacy is appropriate or acceptable in various circumstances (Muir and Howard, 1999; Kapuscinski and Hallerman, 1990; Devlin and Donaldson, 1992). Again, these are issues that the CVM generally has not had to address in the past.

The committee's concern about legal and technical capacity is not limited to the CVM. It is not clear to the committee whether any of the agencies with a possible regulatory role in overseeing the environmental impacts of animal biotechnology has a clear and adequate mandate and the necessary scientific and technical expertise to address these potential impacts. The committee has not made an exhaustive inquiry on this point and has drawn no conclusions, but it believes that the legal and technical capacity of the agencies in the environmental area is a significant concern.

SOCIOECONOMIC, CULTURAL, RELIGIOUS, AND ETHICAL FACTORS

The commercialization of animal biotechnology will occur in the context of existing agricultural and social systems. This technology has the potential to affect a host of social, economic, religious, cultural, and ethical values and interests inside and outside of the agricultural system. Some of these effects might directly be relevant to the mandates of the regulatory agencies. Many are not directly relevant to regulatory mandates, despite their importance to citizens and society. Experience with biotechnology has taught, however, that, even when the social and economic aspects of the technology are beyond the regulatory jurisdiction of an agency, they can affect the questions regulatory agencies are pressed to address by various groups, and sometimes can dominate the public debate in ways that have unavoidable spillover effects on the regulatory process. For this reason, the committee considers it appropriate to identify some of the potential social implications of animal biotechnology. The committee notes as a concern the need for the regulatory agencies to be clear about the scope and limitation of their mandates to address such matters that do not directly affect health, the environment, and animal welfare. Lack of clarity on which issues are within the regulatory mandate and which need to be addressed in other settings could undermine the ability of the agencies to address health and environmental concerns in a scientifically sound and publicly acceptable manner.

Industry Structure and Indirect Health, Animal Welfare, and Environmental Effects

An important economic issue surrounding both plant and animal biotechnology is whether the technology is scale-sensitive—that is, whether it is equally viable economically for both small-scale and large-scale farmers, or whether it favors large-scale or "industrial" styles of agriculture. This question is posed in light of the well-documented trend toward concentration (fewer but larger farms) in U.S. agriculture and a concern that more intensive approaches to plant and animal production can have their own health and environmental impacts.

A large body of scholarly work identifies complex linkages among technologies, their impacts on social systems, and resultant health effects (Barbour, 1993). The "Green Revolution" in agriculture provides many examples throughout the world. The introduction of hybrid seed varieties into high-input, mechanized, industrial farming practices resulted in changes in land use practices and fundamentally changed the social structure of some communities (DeWalt, 1988). In some instances, creation of habitat favorable to mosquito vectors led to increases in the incidence of malaria (Cleaver, 1972;

Sharma, 1999; Middendorf et al., 2000). Because the Green Revolution involved many technical changes, it sometimes is difficult to understand cause-and-effect relationships among the new cultivars, pesticide use, mechanization, land use, and environmental changes, and their social and public health impacts. However, the Green Revolution clearly demonstrates that commercialization of cultivars with relatively simple genetic changes can have major effects on farming practices that ultimately result in environmental and social change (Conway, 1998).

Some have suggested that the use of certain biotechnologies in vertically integrated agricultural operations producing swine and poultry inherently favor a particular kind of large-scale agricultural system under the increasing control of large corporations, with resultant unfavorable economic impacts on smaller-scale farmers and producers (Martinez, 1999). Large-scale agricultural operations might, in turn, have a very different impact on both the natural environment and communities of people than other systems of food production. For example, swine genetically engineered for disease resistance might be raised successfully in increasingly large, crowded feedlots, with resultant impacts on public health and the environment (Cole et al., 2000). In this case, it is difficult to ignore the contribution of the technology to this sequence of events, even though the genetic engineering of swine is not the proximate cause of the impacts.

Concentration in agriculture is the result of many complex economic forces, of which technologic change is just one. The degree to which animal biotechnologies will contribute to a shift from smaller-scale to larger-scale operations, however, sometimes is unclear (Martin, 1991). For example, genetically engineered pigs have been developed to produce phytase, an enzyme that reduces the phosphorus content of manure when the animals' feed contains phytate from plant seeds. The environmental problem(s) posed by phosphorous-rich manure might differ among small- and large-scale agricultural operations. Whether or not genetically engineered pigs able to utilize phytic acid directly in their diet will be equally beneficial to or affordable by small- and large-scale farmers remains to be seen.

These examples suggest that the environmental and social impacts of the shift to larger agricultural operations in some cases might be attributable—at least in part—to the adoption of the genetic technology, though they might not be apparent in an evaluation that focuses narrowly on the direct impacts of the technology. The committee's concern is that there be clarity about whether the regulatory agencies consider it their charge to consider: (1) only the direct health and environmental impacts of a technology, where the technology is the proximate cause of an effect, or (2) also, the social or economic impacts of a technology that can cause an adverse health or environmental impact. The Green Revolution teaches that the scope and size of the social and economic impacts are difficult to predict in the early stages of introducing a technology into the marketplace.

Religious, Spiritual, and Cultural values

Some religious, spiritual, ethnic, or cultural groups prescribe dietary norms or rules that include foods that are to be avoided. These norms or religious traditions might be violated by genetic engineering of animals used as food. A genetically engineered animal might contain a gene or gene product from a prohibited animal, or the mixing of genetic elements from distinct species itself might be prohibited. For example, a human protein derived from biopharming might enter the human food chain if animals are not properly segregated. These techniques might affect the acceptability of the food product to some members of the general public, and have obvious implications for any potential labeling policy (see Box 7.2).

BOX 7.2
Labeling

There has been considerable debate and continuing controversy in the United States and globally about the labeling of foods derived from genetically engineered plants. The committee recognizes the importance of the labeling issue and its potential relevance to animal biotechnology. The current FDA labeling policy requires that foods derived from genetically engineered plants be labeled to inform consumers if there has been a change in the food that would be material to them with respect to safety or nutrition. The committee assumes that a similar policy would apply to foods derived from genetically engineered animals. To date, no genetically engineered plants or food derived from such plants have required labeling under this policy.

Labeling currently is not required in the United States, however, solely to inform consumers that the food was derived from a genetically engineered plant. Some have challenged this policy on the ground that there are reasons—beyond safety or nutrition—for a consumer to want labeling of food derived from genetically engineered plants or animals, including religious, ethical, right-to-know, or simple preference reasons. It could be argued that in the current climate surrounding biotechnology, the fact of genetic engineering is an aspect of the identity of a food derived from a genetically engineered organism. The committee notes, however, that while any one or all of these reasons might provide a legitimate basis in public policy for requiring labeling of biotechnology-derived foods, these are not science-based concerns, and whether they justify labeling is beyond the committee's charge.

In the event labeling is required, however, whether for safety, nutritional, or other reasons, implementation will pose science-based concerns having to do primarily with the availability of simple, reliable test methods to verify whether foods are labeled properly with respect to genetic engineering. The availability of such methods also might be a concern in the event that marketers of food derived from genetically engineered organisms seek to segregate their products from non-genetically engineered food for commercial reasons. The committee acknowledges these technical issues regarding the implementation of labeling and segregation regimens but considers them beyond the scope of its charge.

Ethical Considerations

Ethical considerations might be applicable to the processes involved in biotechnologies as well as the products derived from them. Ethical considerations, of course, are not new to the specific biotechnologies discussed in this report, but the general public often makes ethical distinctions among genetic engineering in plants and animals for biomedical research, for pharmaceutical production, and for food production (Sparks, 1995; Frewer et al., 1997).

Ethical considerations generally are normative and cannot be resolved scientifically. Yet, to ignore them is to assume that science can and should be value-free—an obvious contradiction, since this is a normative assertion in itself (Thompson, 2001). Moreover, as noted above, values can influence both the design of scientific inquiry and the interpretation of data, and certainly can motivate much of the pressure brought to bear on regulatory agencies and other government bodies to address impacts of biotechnology beyond those directly affecting health and the environment.

One view, which focuses on the consequences of applying animal biotechnologies, holds that their ethical significance must derive from the risk to people, animals, and/or the environment (Rollin, 1986; Thompson, 2001). This utilitarian conceptualization sees the technology as directly or indirectly initiating event(s) that are knowable and, to some extent, quantifiable. Through this lens, an ethical analysis would, for example, address the degree of pain and suffering of animals and/or defined risks to human health or the environment, and would draw conclusions based on those consequences. Of course, quantifying pain and suffering or risks associated with hazards about which there is considerable uncertainty remains a significant problem. In fact, some people conclude that because some genetic technologies are characterized by large uncertainties about their consequences, and for high-stakes decisions, it is morally irresponsible to proceed with their application (The Royal Society, 2001).

Some people, however, without regard to the purpose to which the technology is to be put or its consequences, consider genetic engineering of animals fundamentally unethical. This stance might be based on the belief that these technologies violate certain rights or appropriate relationships between humans and nature or God, independent of the consequences of the technologies. This stance also might be based on the conviction that animals have certain rights. Sentience, or the capacity for sensation or feeling, sometimes is used as the quality necessary for moral consideration. Another somewhat related view holds that genetic manipulation of animals for human purposes is disrespectful, and inappropriately interferes fundamentally with animal integrity, dignity, or essential nature. In an address to the Royal Society of Agriculture, Heap (1995) stated, "Programmes which threaten an animal's characteristics and form by restricting its ability to reproduce normally, or which might in the future diminish its behavior or cognition to improve productivity would raise serious intrinsic objections because of their assault on an animal's essential nature." Yet another view focuses on the right of humans to know what they are eating or how their food or pharmaceuticals are being produced. That right—in this view—is independent of biologic risk. In this view, food produced through technologies that some people find "unnatural" or simply "novel" would need to be identified so that consumers could make informed purchasing and dietary decisions. Others argue that if the product has not been altered materially and is deemed safe, it should not be singled out as being different, just as milk produced from cows given growth hormone (BST) was not so labeled. Here, the product, rather than the process that produced the product, is the primary concern.

Unlike a utilitarian approach that considers risks and benefits in the aggregate, a rights-based perspective also looks closely at the distribution of risks and benefits of a technology and its products among individuals. Distributive justice becomes an important consideration.

INTERSECTION OF ETHICS, SCIENCE, AND PUBLIC POLICY

Ethical concerns cannot be resolved completely through scientific debate. Yet, the nature, scope, and direction of scientific research and scientific practice are influenced by ethical considerations. Inasmuch as ethical concerns cannot be separated cleanly from scientific concerns, a strong case can be made that the ethical assumptions underlying a research initiative or the application of a technology should be made explicit. The committee acknowledges that each regulatory authority will bring its own scope of inquiry and set of ethical assumptions to attempts to address the science-based concerns posed by animal biotechnologies.

A historical review of similar efforts suggests that regulatory agencies are likely to focus almost exclusively on what they believe to be the direct impacts of these technologies on human health, food safety, the environment, and in some cases, animal welfare. How each agency will deal with the scope and degree of scientific uncertainty remains to be seen. The full range of socioeconomic impacts of these technologies, however, though likely to be significant and certainly amenable to study using scientific methods, is unlikely to be examined comprehensively and weighed in regulatory decision-making. Moreover, spiritual and religious considerations are unlikely to be addressed substantively at all. Experience also teaches, however, that the public's interest in these value-laden matters can affect the work of regulatory agencies, as evidenced by the European experience with plant biotechnology.

The committee notes that the technologies discussed in this report are likely to have direct and indirect impacts that will become more apparent over time and will generate considerable debate and uncertainty in some instances.

This report was written as these new technologies continued to emerge. It is, therefore, a "snapshot" of a rapidly evolving and complex field of study. It undoubtedly will require revision as new information becomes available.

References

Abrahams, M. V., and A. Sutterlin. 1999. The foraging and antipredator behavior of growth-enhanced transgenic Atlantic salmon. Animal Behavior 58:933–942.

Allen, S. K., S. L. Downing, and K. K. Chew. 1989. Hatchery manual for producing triploid oysters. Seattle, WA: University of Washington Press.

Altekruse, S. F., M. L. Cohen, and D. L. Swerdlow. 1997. Emerging food-borne diseases. Emerging Infectious Diseases 3:285–293.

Alton, E., U. Griesenbach, and D. M. Geddes. 1998. Milking gene therapy. Nature Science 4:1121–1122.

Amsterdam, A., C. Yoon, M. Allende, T. Becker, K. Kawakami, S. Burgess, N. Gaiano, and N. Hopkins. 1997. Retrovirus-mediated insertional mutagenesis in zebrafish and identification of a molecular marker for embryonic germ cells. Cold Spring Harbor Symposium on Quantitative Biology 62:437–450.

Anderson, R. A., K. A. Feathergill, X. H. Diao, M. D. Cooper, R. Kirkpatrick, B. C. Herold, G. F. Doncel, C. J. Chany, D. P. Waller, W. F. Rencher, and L. J. D. Zaneveld. 2002. Preclinical evaluation of sodium cellulose sulfate (Ushercell) as a contraceptive antimicrobial agent. Journal of Andrology 23:426–438.

Andrews, A. L. May 9, 1997. WTO overrules Europe's ban on U.S. hormone-treated beef. The New York Times.

Aritomi, S., and N. Fujihara. 2000. Production of chicken chimeras by fusing blastodermal cells with electroporation. Asian Journal of Andrology 2(4):271–275.

Ashburner, M., M. A. Hoy, and J. Peloquin. 1998. Transformation of arthropods—Research needs and long term prospects. Insect Molecular Biology 7:201–213.

Ashe, H. J., J. Monks, M. Wijgerde, P. Fraser, and N. J. Proudfoot. 1997. Intergenic transcription and transinduction of the human β-globin locus. Genes and Development 11:2494–2509.

Atkinson, P. W., A. C. Pinkerton, and D. A. O'Brochta. 2001. Genetic transforamtion systems in insects. Annual Review of Entomology 46:317–346.

Banner, M. C. 1995. Report of the Committee to Consider the Ethical Implications of Emerging Technologies in the Breeding of Farm Animals. London: Her Majesty's Stationery Office.

Barbour, I. 1993. Ethics in an Age of Technology: The Gifford Lectures. San Francisco: Harper Collins.

Barnes, F. L. 1999. The effects of the early uterine environment on the subsequent development of embryo and fetus. Theriogenology 53:649–658.

Bauman, D. E. 1999. Bovine somatotropin and lactation: From basic science to commercial application. Domestic Animal Endocrinology 17:101–116.

BBC (British Broadcasting Corporation). 2001. Farm vaccine report launched. BBC News. Available online at *news.bbc.co.uk/hi/English/uk/newsid_1590000/1590461.stm.*

Beardmore, J. A., G. C. Mair, and R. I. Lewis. 2001. Monosex male production in finfish as exemplified by tilapia: Applications, problems, and prospects. Aquaculture 197:283–302.

Beernsten, B. T., A. A. James, and B. M. Christensen. 2000. Genetics of mosquito vector competence. Microbiology and Molecular Biology Reviews 64:115.

Benfey, T. J. 1999. The physiology and behavior of triploid fishes. Reviews in Fisheries Science 7:39–67.

Benson, S. J., B. L. Ruis, A. M. Fadly, and K. F. Conklin. 1998. The unique envelope gene of the subgroup J avian leukosis virus derives from ev/J proviruses, a novel family of avian endogenous viruses. Journal of Virology 72(12):10157–10164.

Berkowitz, R., H. Ilves, W. Y. Lin, K. Eckert, A. Coward, S. Tamaki, G. Veres, and I. Plavec. 2001. Construction and molecular analysis of gene transfer systems derived from bovine immunodeficiency virus. Journal of Virology 75(7):3371–3382.

Berlanga, J., J. Infante, V. Capo, J. de la Fuente, and F. O. Castro. 1993. Characterization of transgenic mice linkages. I. Overexpression of hGH causes the formation of liver intranuclear pseudoinclusion bodies and renal and hepatic injury. Acta Biotechnology 13:361–371.

Bertolini, M. 2002. Physiological and Molecular Mechanisms Underlying the Appearance of the Large Calf Syndrome of In Vitro-Produced Bovine Embryos. Ph.D. dissertation. Davis, CA: University of California.

Betthauser, J., E. Forsberg, M. Augenstein, L. Childs, K. Eilertsen, J. Enos, T. Forsythe, P. Golueke, G. Jurgella, R. Koppang, T. Lesmeister, K. Mallon, G. Mell, P. Misica, M. Pace, M. Pfister-Genskow, N. Strelchenko, G. Voelker, S. Watt, S. Thompson, and M. Bishop. 2000. Production of cloned pigs from in vitro systems. Nature Biotechnology 18(10):1055–1059.

Betts, D. H., V. Bordignon, J. R. Hill, Q. Winger, M. E. Westhusin, L. C. Smith, and W. A. King. 2001. Reprogramming of telomerase activity and rebuilding of telomere length in cloned cattle. Proceedings of the National Academy of Sciences of the United States of America 98:1077–1082.

Bird, A. R., W. J. Croom, B. L. Black, Y. K. Fan, and L. R. Daniel. 1994. Somatotropin transgenic mice have reduced jejunal transport rates. Journal of Nutrition 124:2189–2196.

Bishhoff, R., E. Degryse, F. Perraud, W. Dalemans, D. Ali-Hadji, and D. Thepot. 1992. A 17.6 kb region located upstream of the rabbit WAP gene directs high level of expression of a functional human protein variant in transgenic mouse milk. FEBS Letters 305:265–268.

Blair, C. D., Z. N. Adelman, and K. E. Olsen. 2000. Molecular strategies for interrupting arthropod-borne virus transmission by mosquitos. Clinical Microbiology Reviews 13:651.

Bleck, G. T., B. R. White, D. J. Miller, and M. B. Wheeler. 1998. Production of bovine alpha-lactalbumin in the milk of transgenic pigs. Journal of Animal Science 76:3072–3078.

Bleck, G. T., R. Jimenez-Flores, and R. D. Bremel. 1995. Abnormal properties of milk from transgenic mice expressing bovine beta-casein under control of the bovine alpha-lactalbumin 5' flanking region. International Dairy Journal 5(6):619–632.

Boeke, J. D., and J. S. Stoye. 1997. Retrotransposons, endogenous retroviruses, and the evolution of retroelements. Pp. 343–435 in Retroviruses, J. M. Coffin, S. H. Hughes, and H. E. Varmus, eds. Cold Spring Harbor, NY: Cold Spring Harbor Laboratory Press.

Boland, M. P., and J. F. Roche. 1993. Embryo production: Alternative methods. Molecular Reproduction and Development 36:266–270.

Boneva, R. S., T. M. Folks, and L. E. Chapman. 2001. Infectious disease issues in xenotransplantation. Clinical Microbiology Reviews 14(1):1–14.

Bonifer, C., M. C. Huber, U. Jagle, N. Faust, and A. E. Sippel. 1996. Prerequisites for tissue specific and position independent expression of a gene locus in transgenic mice. Journal of Molecular Medicine 74(11):663–671.

Bordignon, V., H. J. Clarke, and L. C. Smith. 2001. Factors controlling the loss of immunoreactive somatic histon H1 from blastomere nuclei in oocyte cytoplasm: A potential marker of reprogramming. Developmental Biology 233(1):192–203.

Braig, H. R., and G. Yan. 2002. The spread of genetic constructs in natural insect populations. Pp. 251–314 in Genetically Engineered Organisms: Assessing Environmental and Human Health Effects, D. K. Letourneau and B. E. Burrows, eds. Washington, D.C.: CRC Press.

Briggs, R. W., E. U. Green, and T. J. King. 1951. An investigation of the capacity for cleavage and differentiation in Rana pipiens eggs lacking "functional" chromosomes. Journal of Experimental Zoology 116:455–499.

Brink, M. F., M. D. Bishop, and F. R. Pieper. 2000. Developing efficient strategies for the generation of transgenic cattle which produce biopharmaceuticals in milk. Theriogenology 53(1):139–148.

Briskin, M. J., R. Y. Hsu, T. Boggs, J. A. Schultz, W. Rishell, and R. A. Bosselman. 1991. Heritable retroviral transgenes are highly expressed in chickens. Proceedings of the National Academy of Sciences of the United States of America 88(5):1736–1740.

Britt, J. H. 1985. Enhanced reproduction and its economic implications. Journal of Dairy Science 68:1585–1592.

Bronson, S. K., and O. Smithies. 1994. Altering mice by homologous recombination using embryonic stem cells. Journal of Biological Chemistry 269(44):27155–27158.

Broom, D. M. 1993. Assessing the welfare of modified or treated animals. Livestock Production Science 36:39–54.

Brown, F., A. M. Lewis, K. Peden, and P. Krause. 2001. Evolving Scientific and Regulatory Perspectives on Cell Substrates for Vaccine Development. Basel, Germany: Karger.

Browning, M. T., R. D. Schmidt, K. A. Lew, and T. A. Rizvi. 2001. Primate and feline lentivirus vector RNA packaging and propagation by heterologous lentivirus virions. Journal of Virology 75(11):5129–5140.

Bruce, M. E., R. G. Will, J. W. Ironside, I. McConnell, D. Drummond, A. Suttie, L. McCardle, A. Chree, J. Hope, C. Birkett, S. Cousens, H. Fraser, and C. J. Bostock. 1997. Transmissions to mice indicate that "new variant" CJD is caused by the BSE agent. Nature 389:498–501.

Bueler, H., M. Fischer, Y. Lang, H. Bluethmann, H. P. Lipp, S. J. DeArmond, S. B. Prusiner, M. Aguet, and C. Weissmann. 1992. Normal development and behaviour of mice lacking the neuronal cell-surface PrP protein. Nature 356(6370):577–582.

Burdon, T., R. J. Wall, A. Shamay, G. H. Smight, and L. Hennighausen. 1991. Overexpression of an endogenous milk protein gene in trangenic mice is associated with impaired mammary gland alveolar development and a milchlos phenotype. Mechanisms of Development 36:67–74.

Burghardt, G. M., and H. A. Herzog. 1989. Animals, evolution, and ethics. Pp. 129–151 in Perceptions of Animals in American Culture, R. J. Hoage, ed. Washington, DC: Smithsonian Institution Press.

Burrin, D. G. 1997. Is milk-borne insulin-like growth factor-I essential for neonatal development? Journal of Nutrition 127:975S–979S.

Bye, V. J., and R. F. Lincoln. 1986. Commercial methods for the control of sexual maturation in rainbow trout (*Salmo gairdneri* R.) Aquaculture 57:299–309.

Byrne, G. W., K. R. McCurry, M. J. Martin, S. M. McClellan, J. L. Platt, and J. S. Logan. 1997. Transgenic pigs expressing human CD59 and decay-accelerating factor produce an intrinsic barrier to complement-mediated damage. Transplantation 63:19–155.

Calvo, G. W., M. W. Luckenbach, S. K. Allen, Jr., and E. M. Burreson. 2001. Comparative field study of Crassostrea gigas and Crassostrea virginica in relation to salinity in Virginia. Journal of Shellfish Research 20:221–229.

Campbell, K. H. 1999. Nuclear equivalence, nuclear transfer, and the cell cycle. Cloning 1:3–15.

Campbell, K. H., J. McWhir, W. A. Ritchie, and I. Wilmut. 1996. Sheep cloned by nuclear transfer from a cultured cell line. Nature 380(6569):64–66.

Campton, D. E. 1987. Natural hybridization and introgression in fishes: Methods of detection and genetic interpretations. Pp. 161–192 in Population Genetics and Fishery Management, N. Ryman and F. Utter, eds. Seattle: University of Washington Press.

Capy, P., D. Anxolabehere, and T. Langin. 1994. The strange phylogenies of transposable elements—Are horizontal transfers the only explanation? Trends in Genetics 10:7–12.

Caron, D. M. 2002. Africanized honey bees in the Americas. American Bee Journal 142:327–328.

Carr, J. W., J. M. Anderson, F. G. Whoriskey, and T. Dilworth. 1997. The occurrence and spawning of cultured Atlantic salmon (*Salmo salar*) in a Canadian river. International Council for the Exploration of the Seas Journal of Marine Science 54:1064–1073.

Castilla, J., B. Pintado, I. Sola, J. M. Sanchez-Morgado, and L. Enjuanes. 1998. Engineering passive immunity in transgenic mice secreting virus-neutralizing antibodies in milk. Nature Biotechnology 16:349–354.

Chakraborty, A. K., M. A. Zink, and C. P. Hodgson. 1994. Transmission of endogenous VL30 retrotransposons by helper cells used in gene therapy. Cancer Gene Therapy 1(2):113–118.

Chan, A. W., E. J. Homan, L. U. Ballou, J. C. Burns, and R. D. Bremel. 1998. Transgenic cattle produced by reverse-transcribed gene transfer in oocytes. Proceedings of the National Academy of Sciences of the United States of America 95(24):14028–14033.

Chen T. T., C. Lin, M. Shamblott, J. K. Lu, and K. Knight. 1994. Transgenic fish and aquaculture. Pp. 324–331 in Proceedings, 5th World Congress on Genetics Applied to Livestock Production, C. Smith, J. S. Gavora, B. Benkel, J. Chesnais, W. Fairfull, J. P. Gibson, B. W. Kennedy, and E. B. Burnside, eds. Ontario: University of Guelph Press.

Chiu, C. H., T. L. Wu, L. H. Su, C. S. Chu, J. H. Chia, A. J. Kuo, M. S. Chien, and T. Y. Lin. 2002. The Emergence in Taiwan of fluoroquinolone resistance in Salmonella enterica serotype choleraesuis. New England Journal of Medicine 346(6):413–419.

Cibelli, J. B., K. H. Campbell, G. E. Seidel, M. D. West, and R. P. Lanza. 2002. The health profile of cloned animals. Nature Biotechnology 20(1):13–14.

Cibelli, J. B., S. L. Stice, P. J. Golueke, J. J. Kane, J. Jerry, C. Blackwell, F. A. Ponce de Leon, and J. M. Robl. 1998. Transgenic bovine chimeric offspring produced from somatic cell-derived stem-like cells. Nature Biotechnology 16(7):642–646.

Cleaver H. M. 1972. The contradictions of the green revolution. Monthly Review 24:80–111.

Cole D., L. Todd, and S. Wing. 2000. Concentrated swine feeding operations and public health: A review of occupational and community health effects. Environmental Health Perspectives 108(8):685–699.

Colman, A. 1996. Production of proteins in the milk of transgenic livestock: Problems, solutions, and successes. American Journal of Clinical Nutrition 63:639S–645S.

Conway, G. 1998. The Doubly Green Revolution: Food for All in the 21st Century. Ithaca, NY: Comstock.

Cook, J. T., M. A. McNiven, G. F. Richardson, and A. M. Sutterlin. 2000. Growth rate, body composition, and food digestibility/conversion of growth-enhanced Atlantic salmon (*Salmo salar*). Aquaculture 188:15–32.

Cowan P. J., A. Aminian, H. Barlow, A. A. Brown, C. G. Chen, N. Fisicaro, D. M. Francis, D. J. Goodman, W. Han, M. Kurek, M. B. Nottle, M. J. Pearse, E. Salvaris T. A. Shinkel, G. V. Sainsbury, A. B. Stewart, and A. J. A'pice. 2000. Renal xenografts from triple-transgenic pigs are not hyperacutely rejected but cause coagulopathy in non-immunosuppressed baboons. Transplantation 69:2504–2515.

Cozzi, E., A. W. Tucker, G. A. Langford, G. Pino-Chavez, L. Wright, M. J. O'Connell, V. J. Young, R. Lancaster, M. McLaughlin, K. Hunt, M. C. Bordin, and D. J. White. 1997. Characterization of pigs transgenic for human decay-accelerating factor. Transplantation 64:1383–1392.

Cozzi, E., and D. J. G. White. 1995. The generation of transgenic pigs as potential organ donors for humans. Nature Medicine 1:964–966.

Crittenden, L. B., and D. W. Salter. 1990. Expression of retroviral genes in transgenic chickens. Journal of Reproduction and Fertility 41:163S–171S.

Cummings, M. P. 1994. Transmission patterns of eukaryotic transposable elements— Arguments for and against horizontal transfer. Trends in Ecology and Evolution 9:141–145.

Cummins, J. M. 2001. Cytoplasmic inheritance and its implications for animal biotechnology. Theriogenology 55:1381–1399.

Curran, M. A., S. M. Kaiser, P. L. Achacoso, and G. P. Nolan. 2000. Efficient transduction of nondividing cells by optimized feline immunodeficiency virus vectors. Molecular Therapy 1(1):31–38.

Dai, Y., T. D. Vaught, J. Boone, S. H. Chen, C. J. Phelps, S. Ball, J. A. Monahan, P. M. Jobst, K. J. McCreath, A. E. Lamborn, J. L. Cowell-Lucero, K. D. Wells, A. Colman, I. A. Polejaeva, and D. L. Ayares. 2002. Targeted disruption of the alpha 1,3-galactosyltransferase gene in cloned pigs. Nature Biotechnology 20(3):251–255.

Dangel, J. R., P. T. Macey, and F. C .Withler. 1973. An annotated bibliography of interspecific hybridization of fishes of the subfamily Salmoninae. U.S. Department of Commerce, National Oceanic and Atmospheric Administration (NOAA) Technical Memorandum WNMFSFC-1, U.S.

Daniels, R., V. Hall, and A. O. Trounson. 2000. Analysis of gene transcription in bovine nuclear trasfer embryos reconstructed with granulose cell nuclei. Biology of Reproduction 63:1034–1040.

DANMAP (Danish Integrated Antimicrobial Resistance Monitoring and Research Program). 2000. Consumption of antimicrobial agents and occurrence of antimicrobial resistance in bacteria from food animals, foods and humans in Denmark. Copenhagen: Danish Integrated Antimicrobial Resistance Monitoring and Research Program. Available online at http//www.svs.dk/dk/Publikationer/Danmap/Danmap %202000.pdf.

De Sousa, P. A., Q. Winger, J. Hall, K. Jones, A. Watson, and M. Westhusin. 1999. Reprogramming of fibroblast nuclei after transfer into bovine oocytes. Cloning 1:63–69.

De Sousa, P. A., T. King, L. Harkness, L. E. Young, S. K. Walker, and I. Wilmut. 2001. Evaluation of gestational deficiencies in cloned sheep fetuses and placentae. Biology of Reproduction 65:23–30.

Dekkers, J. C. M., and J. A. M. Van Arendonk. 1998. Optimum selection for quantitative traits with information on an identified locus in outbred populations. Genetical Research 71:257–275.

Dekkers, J. C., and F. Hospital. 2002. The use of molecular genetics in the improvement of agricultural populations. Nature Reviews Genetics 3:22–32.

Dennis, M. B., Jr. 2002. Welfare issues of genetically modified animals. ILAR (Institute for Laboratory Animal Research) Journal 43(2):100–109.

Devinoy, E., D. Thepot, M. G. Stinnakre, M. L. Fontaine, H. Grabowski, and C. Puissant. 1994. High level production of human growth hormone in the milk of transgenic mice: The upstream region of the rabbit whey acidic protein (WAP) gene targets transgene expression in the mammary gland. Transgenic Research 3:79–89.

Devlin, R. H., and E. M. Donaldson. 1992. Containment of genetically altered fish with emphasis on salmonids. Pp. 229–265 in Transgenic fish, C. L. Hew and G. L. Fletcher, eds. Co. River Edge, NJ: World Scientific Publishing.

Devlin, R. H., C. A. Biagi, T. Y. Yesaki, D. E. Smailus, and J. C. Byatt. 2001. A growth-hormone transgene boosts the size of wild but not domesticated trout. Nature 409:781–782.

Devlin, R. H., J. I. Johnsson, D. E. Smailus, C. A. Biagi, E. Jonsson, and B. T. Bjornsson. 1999. Increased ability to compete for food by growth hormone-transgenic coho salmon Oncorhynchus kisutch (Walbaum). Aquaculture Research 30:479–482.

Devlin, R. H., T. Y. Yesaki, C. A. Biagi, E. M. Donaldson, P. Swanson, and W. K. Chan. 1994. Extraordinary salmon growth. Nature 371:209–210.

Devlin, R. H., T. Y. Yesaki, E. Donaldson, S. Du, and C. Hew. 1995a. Production of germline transgenic pacific salmonids with dramatically increased growth-performance. Canadian Journal of Fisheries and Aquatic Science 52:1376–1384.

Devlin, R. H., T. Y. Yesaki, E. M. Donaldson, and C. L. Hew. 1995b. Transmission and phenotypic effects of an antifreeze/GH gene construct in coho salmon (Oncorhynchus kisutch). Aquaculture 137:161–169.

DeWalt, B. 1988. Halfway there: Social science in agriculture development and the social science of agriculture development. Human Organization 47(4):343–353.

DHHS (Department of Health and Human Services). 2001. PHS (Public Health Service) guideline on infectious disease issues in xenotransplantation. Available online at *http://www.fda.gov/cber/gdlns/xenophs0101.pdf*.

Diamond, J. 1999. Guns, Germs, and Steel: The Fates of Human Societies. New York: WW Norton & Company.

Diles, J. J. B., R. D. Green, H. H. Shepherd, G. L. Mathiews, L. J. Hughes, and M. F. Miller. 1996. Relationships between body measurements obtained on yearling Brangus bulls and measures of carcass merit obtained from their steer clone-mates. The Professional Animal Scientist 12:244–249.

Dodgson, J. B., S. H. Hughes, and M. J. Federspiel. 1999. Soluble forms of the subgroup A avian leukosis virus [ALV(A)] receptor tva significantly inhibit ALV(A) infection in vitro and in vivo. Journal of Virology 73:10051–10060.

Dove, A. 2000. Milking the genome for profit. Nature Biotechnology 18:1045–1050.

Doyle, E. 2000. Human safety of hormone implants used to promote growth in cattle: A review of the scientific literature. Madison, WI: FRI (Food Research Institute) Briefings.

Driancourt, M. A. 2001. Regulation of ovarian follicular dynamics in farm animals. Implications for manipulation of reproduction. Theriogenology 55:1211–1239.

Droge, M., A. Puhler, and W. Selbitschka. 1998. Horizontal gene transfer as a biosafety issue: A natural phenomenon of public concern. Journal of Biotechnology 64:75–90.

Dunham, R. A. 1994. Predator avoidance, spawning, and foraging ability of transgenic catfish. In The 1994 Risk Assessment Research Symposium. Available online at *http://www.nbiap.vt.edu/brarg/brasym94/dunham.htm*.

Dunham, R. A. 1996. Results of early pond-based studies of risk assessment regarding aquatic GMOs. In The 126th Annual Meeting of the American Fisheries Society, Dearborn, MI, August 26–29, 1996. Abstract #381.

Dunhan, R. A., G. W. Warr, A. Nichols, P. L. Duncan, B. Argue, D. Middleton, and H. Kucuktas. 2002. Enhanced bacterial disease resistance of transgenic channel catfish Ictalurus punctatus possessing cecropin genes. Marine Biotechnology 4:338–344.

Ebert, K. M., and J. E. S. Schindler. 1993. Transgenic farm animals: Progress report. Theriogenology 39:121–135.

Ebert, K. M., P. DiTullio, C. A. Barry, J. E. Schindler, S. L. Ayares, and T. E. Smith. 1994. Induction of human tissue plasminogen activator in the mammary gland of transgenic goats. Biotechnology 12:699–702.

Evans, M. J., C. Gurer, J. D. Loike, I. Wilmut, A. E. Schnieke, and E. A. Schon. 1999. Mitochondrial DNA genotypes in nuclear transfer-derived cloned sheep. Nature Genetics 23:90–93.

Eyestone, W. H. 1994. Challenges and progress in the production of transgenic cattle. Reproduction, Fertility, and Development 6:647–652.

Eyestone, W. H. 1999. Production of transgenic cattle expressing a recombinant protein in milk. Pp. 177–192 in Transgenic Animals in Agriculture, J. D. Murray, G. B. Anderson, A. M. Oberbauer, and M. M. McGloughlin eds. Wallingford, UK: CABI International.

Eyestone, W. H., and K. H. S Campbell. 1999. Nuclear transfer from somatic cells: Applications in farm animal species. Journal of Reproduction and Fertility Supplement 54:489–497.

FAO (Food and Agriculture Organization of the United Nations). 2001. Evaluation of allergenicity of genetically modified foods. Pp. 1–27 in Joint report by FAO and World Health Organization.

FAO (Food and Agriculture Organization of the United Nations). 2000. World aquaculture production by species groups. Available online at *ftp://ftp.fao.org/fi/stat/summ_00/b-_table.pdf.*

Farin, P. W., and C. E. Farin. 1995. Transfer of bovine embryos produced in vivo and in vitro: Survival and fetal development. Biology of Reproduction 52:676–682.

Farrell, A. P., W. Bennett, and R. H. Devlin. 1997. Growth-enhanced transgenic salmon can be inferior swimmers. Canadian Journal of Zoology 75:335–337.

FASS (Federation of Animal Science Societies). 1999. Guide for the Care and Use of Agricultural Animals in Agricultural Research and Teaching. Savoy, IL: Federation of Animal Science Societies.

FDA (Food and Drug Administration). 1997. Substances prohibited from use in animal food or feed: animal proteins prohibited in ruminant feed. 21 CFR, Part 589. Federal Register 62(108)130935–30978.

FDA (Food and Drug Administration). 2000. Food and Drug Administration pesticide program. In the Center for Food Safety and Applied Nutrition, Food and Drug Administration. Available online at *http://vm.cfsan.fda.gov/~dms/pes99rep.html #suminclevel.*

Fey, P. D., T. J. Safranek, M. E. Rupp, E. F. Dunne, E. Ribot, P. C. Iwen, P. A. Bradford, F. J. Angulo, and S. H. Hinrichs. 2000. Ceftriaxone-resistant Salmonella infection acquired by a child from cattle. New England Journal of Medicine 342:1242–1249.

First, N. L. 1991. New advances in reproductive biology of gametes and embryos. Pp. 1–21 in Animal Applications of Research in Mammalian Development: Current communications in Cell and Molecular Biology, R. A. Petersen, A. McLaren, and N. L. First, eds. Cold Spring Harbor, NY: Cold Spring Harbor Laboratory Press.

Fiske, P., and R. A. Lund. 1999. Escapees of reared salmon in coastal and riverine fisheries in the period 1989–1998. NINA Offdragsmelding 603:1–23 (in Norwegian with an English abstract).

Fletcher, G. L., A. Alderson, E. A. Chin-Dixon, M. A. Shears, S. V. Goddard, and C. L. Hew. 2000. Transgenic fish for sustainable aquaculture. Pp. 193–201 in Sustainable Aquaculture, N. Svennevig, H. Reinertsen, and M. New. Rotterdam, Netherlands: AA. Bolkema.

Fletcher, G. L., P. L. Davies, and C. L. Hew. 1992. Genetic engineering of freeze-resistant Atlantic salmon. Pp. 190–208 in Transgenic fish, C. L Hew and G. L. Fletcher, eds. River Edge, NJ: World Scientific Publishing.

Foote, R. H. 1996. Dairy cattle reproductive physiology research and management—Past progress and future prospects. Journal of Dairy Science 79:980–990.

Foster, E. M. 1982. Is there a food safety crisis? Food Technology 36(8):82–83.

Fox, J. L. 2001. EPA reevaluates StarLink license. Nature Biotechnology 19(1):11.

Fraser, D. 1999. Animal ethics and animal welfare science: Bridging the two cultures. Applied Animal Behaviour Science 65:171–189.

Fraser, D., D. M. Weary, E. A. Pajor, and B. N. Milligan. 1997. A scientific conception of animal welfare that reflects ethical concerns. Animal Welfare 6(3):187–205.

Frewer, L., C. Howard, and R. Shepherd. 1997. Public concerns about general and specific applications of genetic engineering: Risk, benefit, and ethics. Science, Technology, and Human Values. 22:98–124.

Fulka, J., Jr., P. Loi, S. Ledda, R. M. Moor, and J. Fulka. 2001. Nucleus transfer in mammals: how the oocyte cytoplasm modifies the transferred nucleus. Theriogenology 55(6):1373–1380.

Funtowicz, S., and J. Ravetz. 1992. Three types of risk assessment and the emergence of post-normal science. In Social Theories of Risk, S. Krimsky and D. Golding, eds. Westport, CT: Praeger Press.

Galili, U. 2001. The alpha-gal epitope (Galalpha1-3Galbeta1-4GlcNAc-R) in xenotransplantation. Biochimie 83(7):557–563.

Galli, C., and G. Lazzari. 1996. Practical aspects of IVM/IVF in cattle. Animal Reproductive Science 42:371–379.

Garry, F. B., R. Adams, J. P. McCann, and K. G. Odde. 1996. Postnatal characteristics of calves produced by nuclear transfer cloning. Theriogenology 45:141–152.

Gendel, S. M. 1998a. Sequence databases for assessing the potential allergenicity of proteins used in transgenic foods. Advanced Food and Nutrition Research 42:63–92.

Gendel, S. M. 1998b. The use of amino acid sequence alignments to assess potential allergenicity of proteins used in genetically modified foods. Advanced Food and Nutrition Research 42:45–62.

Georges, M. 2001. Recent progress in livestock genomics and potential impact on breeding programs. Theriogenology 55:15–21.

Gerardi, M. H., and J. K. Grimm. 1979. The History, Biology, Damage, and Control of the Gypsy Moth, *Porthetria dispar*. Cranberry, NJ: Associated University Press.

Gibbons, J., S. Arat, J. Rzucidlo, K. Miyoshi, R. Waltenburg, D. Respess, A. Venable, and S. Stice. 2002. Enhanced survivability of cloned calves derived from roscovitine-treated adult somatic cells. Biology of Reproduction 66:895–900.

Golovan S. P., R. G. Meidinger, A. Ajakaiye, M. Cottrill, M. Z. Wiederkehr, D. J. Barney, C. Plante, P. W. Pollard, M. Z. Fan, M. A. Hayes, J. Laursen, J. P. Hjorth, R. R. Hacker, J. P. Phillips, and C. W. Forsberg. 2001a. Pigs expressing salivary phytase produce low-phosphorus manure. Nature Biotechnology 19:741–745.

Golovan S. P., M. A. Hayes, J. P. Phillips, and C. W. Forsberg. 2001b. Transgenic mice expressing bacterial phytase as a model for phosphorus pollution control. Nature Biotechnology 19:429–433.

Gorbach, S. L. 2001. Antimicrobial use in animal feed—Time to stop. New England Journal of Medicine 345:1161–1166.

Gordon, J. W., and F. H. Ruddle. 1985. DNA-mediated genetic transformation of mouse embryos and bone marrow: A review. Gene 33(2):121–136.

Grabowski, H., D. Le Bars, N. Chene, J. Attal, R. Malienou-Ngassa, C. Puissant, and L. M. Houdebine. 1991. Rabbit whey acidic protein concentration in milk, serum, mammary gland extract, and culture medium. Journal of Dairy Science 74:4143–4150.

Grandin, T. 1993. Livestock Handling and Transport. Wallingford, United Kingdom: CABI International.

Grandin, T., and M. J. Deesing. 1998. Genetics and behavior during handling and restraint. Pp. 113–144 in Genetics and the Behavior of Domestic Animals, T. Grandin, ed. San Diego: Academic Press.

Gray, L. E. Jr., C. Lambright, V. Wilson, J. Ostby, L. J. Guillette, and E. Wilson. 2001. In vivo and in vitro androgenic effects of beta trenbolone, a potential feed lot contaminant. Society of Environmental Toxicology and Chemistry (SETAC) 22nd Annual Meeting, Abstract PH070.

Groot C., and L. Margolis. 1991. Pacific salmon life histories. Vancouver: University of British Columbia Press.

Gross, M. L., J. F. Schneider, N. Moav, B. Moav, C. Alvarez, S. H. Myster, Z. Liu, E. M. Hallerman, P. B. Hackett, K. S. Guise, A. J. Faras, and A. R. Kapuscinski. 1992. Molecular analysis and growth evaluation of northern pike (*Esox lucius*) microinjected with growth hormone genes. Aquaculture 103:253–273.

Hagstad, H. V., and W. T. Hubbert. 1981. Food Quality Control: A syllabus for veterinary students. Ames: The Iowa State University Press.

Hale, E. B. 1969. Domestication and the evolution of behavior. Pp. 22–24 in The Behavior of Domestic Animals, E. S. E. Hafez, ed. Baltimore: Williams and Wilkins.

Halford, N. G., and P. R. Shewry. 2000. Genetically modified crops: Methodology, benefits, regulation, and public concerns. British Medical Bulletin 56(1):62–73.

Hallerman, E. M., and A. R. Kapuscinski. 1992a. Ecological and regulatory uncertainties associated with transgenic fish. Pp. 209–228 in Transgenic Fish, C. L. Hew and G. L. Fletcher, eds. Singapore: World Science Publishing Company.

Hallerman, E. M., and A. R. Kapuscinski. 1992b. Ecological implications of using transgenic fishes in aquaculture. International Council for the Exploration of the Seas Marine Science Symposium 194:56–66.

Hallerman, E. M., and A. R. Kapuscinski. 1993. Potential impacts of transgenic and genetically manipulated fish on wild populations: Addressing the uncertainties through field testing. Pp. 93–112 in Genetic Conservation of Salmonid Fishes, J. G. Cloud and G. H. Thorgaard, eds. New York: Plenum Press.

Hallerman, E. M., J. F. Schneider, M. Gross, Z. Liu, S. J. Yoon, L. He, P. B. Hackett, A. J. Faras, A. R. Kapuscinski, and K. S. Guise. 1990. Gene expression promoted by the RSV long terminal repeat element in transgenic goldfish. Animal Biotechnology 1:79–93.

Hamada, M., Y. Kido, M. Himberg, J. D. Reist, C. Ying, M. Hasegawa, and N. Okada. 1997. A newly isolated family of short interspersed repetitive elements (SINEs) in coregonid fishes (whitefish) with sequences that are almost identical to those of the SmaI family of repeats: Possible evidence for the horizontal trasnfer of SINEs. Genetics 146:355–367.

Hammer, R. E., R. L. Brinster, and R. D. Palmiter. 1985. Use of gene transfer to increase animal growth. Cold Spring Harbor Symposia on Quantitative Biology 50:379–387.

Handler, A. M. 2001. A current perspective on insect gene transformation. Insect Biochemistry and Molecular Biology 31(2):111–128.

Hansen, L. P., T. Håstein, G. Nævdal, R. L. Saunders, and J. E. Thorpe. 1991. Interactions between cultured and wild Atlantic salmon. Aquaculture 98(1/3):1–324.

Harlow, S. 2002. Where have all the heifers gone? Dairy Business Communications. Available online at http://www.dairybusiness.com/northeast/Feb02/Feb02NEDBp16.htm.

Hartl, D. L., A. R. Lohe, and E. R. Lozovskaya. 1997. Regulation of the transposable element mariner. Genetica 100:177–184.

Harvey, A. J., G. Speksnijder, L. R. Baugh, J. A. Morris, and R. Ivarie. 2002. Expresion of exogenous protein in the egg white of transgenic chickens. Nature Biotechnology 19:396–399.

Heap, R. B. 1995. Agriculture and bioethics—Harmony or discord? Journal of the Royal Agricultural Society of England 156:69–78.

Hedrick, P. W. 2000. Genetics of Populations, 2nd edition. Sudbury, MA: Jones & Bartlett.

Hedrick, P. W. 2001. Invasion of transgenes from salmon or other genetically modified organisms into natural populations. Canadian Journal of Fisheries and Aquatic Science 58:841–844.

Hefle, S. L., J. A. Nordlee, and S. L. Taylor. 1996. Allergenic foods. Critical Reviews in Food Science and Nutrition 36:69S–89S.

Heitzman, R. J. 1976. The effectiveness of anabolic agents in increasing rate of growth in farm animals: Report in experiments in cattle. Environmental Quality and Safety 5(Supplement):89098.

Henikoff, S. 1998. Conspiracy of silence among repeated transgenes. Bioessays 20(7):532–535.

Hennighausen, L., R. McKnight, T. Burdon, M. Baik, R. J. Wall, and G. H. Smith. 1994. Whey acidic protein extrinsically expressed from the mouse mammary tumor virus long terminal repeat results in hyperplasia of the coagulation gland epithelium and impaired mammary gland development. Cell Growth Differentiation 5:607–613.

Henricks, D. M., S. L. Gray, J. J. Owenby, and B. R. Lackey. 2001. Residues from anabolic preparations after good veterinary practice. APMIS 109(4):173–183.

Her Majesty's Government. 2000. Home office code of practice for the housing and care of pigs intended for use as xenotransplant source animals. Available online at http://www.homeoffice.gov.uk/animalsinsp/reference/codes_of_practice/xenopig.pdf.

Herman, H. A. 1981. Improving Cattle by the Millions: NAAB and the Development and Worldwide Application of Artificial Insemination. Columbia: University of Missouri Press.

Herring, C., G. Quinn, R. Bower, N. Parsons, N. A. Logan, A. Brawley, K. Elsome, A. Whittam, X. M. Fernandez-Suarez, D. Cunningham, D. Onions, G. Langford, and L. Scobie. 2001. Mapping full-length porcine endogenous retroviruses in a large white pig. Journal of Virology 75(24):12252–12265.

Heyman, Y., X. Vignon, P. Chesne, D. Le Bourhis, J. Marchal, and J. P. Renard. 1998. Cloning in cattle: From embryo splitting to somatic nuclear transfer. Reproduction Nutrition Development 38:595–603.

Hill, J. R., A. J. Roussel, J. B. Cibelli, J. F. Edwards, N. L. Hooper, M. W. Miller, J. A. Thompson, C. R. Loonery, M. E. Westhusin, J. M. Robl, and S. L. Stice. 1999. Clinical and pathological features of cloned transgenic calves and fetuses (13 case studies). Theriogenology 51:1451–1465.

Hill, J. R., R. C. Burghardt, K. Jones, C. R. Long, C. R. Looney, T. Shin, T. E. Spencer, J. A. Thompson, Q. A. Winger, and M. E. Westhusin. 2000. Evidence for placental abnormality as the major cause of mortality in first-trimester somatic cell cloned fetuses. Biology of Reproduction 63:1787–1794.

Hindar, K., N. Ryman, and F. Utter. 1991. Genetic effects of cultured fish on natural fish populations. Canadian Journal of Fisheries and Aquatic Science 48:945–957.

Hoekstra, H. E., J. M. Hoekstra, D. Berrigan, S. N. Vignieri, A. Hoang, C. E. Hill, P. Berreli, and J. G. Kingsolver. 2001. Strength and tempo of directional selection in the wild. Proceedings of the National Academy of Sciences of the United States of America 98:9157–9160.

Hone, J. 2002. Feral pigs in Namadgi National Park, Australia: Dynamics, impacts and management. Biological Conservation 105:231–242.

Houck, M. A., J. B. Clark, K. R. Peterson, and M. G. Kidwell. 1991. Possible horizontal transfer of Drosophila genes by the mite Proctolaelaps regalis. Science 253:1125–1128.

Houdebine, L. M. 2000. Transgenic animal bioreactors. Transgenic Research 9:305–320.

Howard, R. D., R. S. Martens, S. A. Innes, J. M. Drnevich, and J. Hale. 1998. Mate choice and mate competition influence male body size in Japanese medaka. Animal Behaviour 55:1151–1163.

Hoy, M. A. 1992a. Commentary: Biological control of arthropods: Genetic engineering and environmental risks. Biological Control 2:166–170.

Hoy, M. A. 1992b. Criteria for release of genetically improved phytoseiids: An examination of the risks associated with release of biological control agents. Experimental and Applied Acarology 14:393–416.

Hoy, M. A. 1995. Impact of risk analyses on pest management programs employing transgenic arthropods. Parasitology Today 11:229–232.

Hoy, M. A. 1997. Laboratory containment of transgenic arthropods. American Entomology 43:206–256.

Hoy, M. A. 2000. Transgenic arthropods for pest management programs risk and realties. Experimental and Applied Acarology 24:463–495.

Hoy, M. A. 2001. Transgenic Insects: Potential Risk Assessment Issues. Presentation to National Academies of Science Public Workshop, Defining Science-based Concerns Associated with Products of Animal Biotechnology. 27 Nov. Washington, DC. Available online at *http://video.nationalacademies.org/ramgen/dels/112701_j.rm.*

Ito, J., A. Ghosh, L. A. Moreira, E. A. Wimmer, and M. Jacobs-Lorena. 2002. Transgenic anopheline mosquitoes impaired in transmission of a malaria parasite. Nature 417:452–455.

Ivics, Z., P. B. Hackett, R. H. Plasterk, and Z. Izsvak. 1997. Molecular reconstruction of Sleeping Beauty, a Tc1-like transposon from fish, and its transposition in human cells. Cell 91(4):501–510.

Izsvak, Z., Z. Ivics, and R. H. Plasterk. 2000. Sleeping Beauty, a wide host-range transposon vector for genetic transformation in Journal of Molecular Biology 302 (1):93–102.

Jaenisch, R. 1997. DNA methylation and imprinting: Why bother? Trends in Genetics 13(8):323–329.

Jaenisch, R., and I. Wilmut. 2001. Developmental biology. Don't clone humans! Science 291:2552.

Jahner, D., and R. Jaenisch. 1985. Chromosomal position and specific demethylation in enhancer sequences of germline-transmitted retroviral genomes during mouse development. Molecular and Cellular Biology 5:2212–2220.

Jahner, D., K. Haase, R. Mulligan, and R. Jaenisch. 1985. Insertion of the bacterial gpt gene into the germ line of mice by retroviral infection. Proceedings of the National Academy of Sciences of the United States of America 82:6927–6931.

Jarvi, T. 1990. The effects of male dominance, secondary sexual characteristics and female mate choice on the mating success of male Atlantic salmon *Salmo salar*. Ethology 84:123–132.

Jeffries, C. 1974. Quantitative stability and digraphics in model ecosystems. Ecology 55:1414–1419.

Jhappan, C., A. G. Geiser, E. C. Kordon, D. Bagheri, L. Hennighausen, A. B. Roberts, G. H. Smith, and G. Merlino. 1993. Targeting expression of a transforming growth factor b1 transgene to the pregnant mammary gland inhibits alveolar development and lactation. European Molecular Biology Organization Journal 12:1835–1845.

Jia, X., A. Patrzykat, R. H. Devlin, P. A. Ackerman, G. K. Iwama, and R. E. Hancock. 2000. Antimicrobial peptides protect coho salmon from *Vibrio anguillarum* infections. Applied and Environmental Microbiology 66:1928–1932.

Johnson, L. A. 2000. Sexing mammalian sperm for production of offspring: The state-of-the-art. Animal Reproduction Science 60–61:93–107.

Johnsson, J. I., E. Jönsson, E. Petersson, T. Jarvi, and B. T. Björnsson. 2000. Fitness-related effects of growth investment in brown trout under field and hatchery conditions. Journal of Fish Biology 57:326.

Johnsson, J. I., E. Petersson, E. Jonsson, and B. T. Bjornsson. 1999. Growth hormone-induced effects on mortality, energy status, and growth: A field study on brown trout (*Salmo trutta*). Functional Ecology 13:514–522.

Johnstone, R., and A. F. Youngson. 1984. The progeny of sex-inverted female Atlantic salmon (*Salmo salar* L.) Aquaculture 37:179–182.

Jones, J. W. 1959. The Salmon. London: Collins.

Jordan, I. K., L. V. Matunina, and J. F. McDonald. 1999. Evidence for the recent horizontal transfer of long terminal repeat retrotransposon. Proceedings of the National Academy of Sciences of the United States of America 96:12621–12625.

Judge, L. J., R. J. Erskine, and P. C. Bartlett. 1997. Recombinant bovine somatotropin and clinical mastitis: Incidence, discarded milk following therapy, and culling. Journal of Dairy Science 80:3212–3218.

Juskevich, J. C., and C. G. Guyer. 1990. Bovine growth hormone: Human food safety evaluation. Science 249:875–884.

Kang, Y. K., D. B. Koo, J. S. Park, Y. H. Choi, H. N. Kim, W. K. Chang, K. K. Lee, and Y. M. Han. 2001. Typical demethylation events in cloned pig embryos: Clues on species-specific differences in epigenetic reprogramming of a cloned donor genome. Journal of Biological Chemistry 276(43):39980–39984.

Kaplan, D. L. 2002. Spiderless spider webs. Nature Biotechnology 20:239–240.

Kapuscinski, A. R. 2002. Controversies in designing useful ecological assessments of genetically engineered organisms. Pp. 385–415 in Genetically Engineered Organisms: Assessing Environmental and Human Health Effects, D. Letourneau and B. Burrows, eds. Boca Raton, FL: CRC Press.

Kapuscinski, A. R., and E. M. Hallerman. 1990. Transgenic fish and public policy: Anticipating environmental impacts of transgenic fish. Fisheries 15:2–11.

Kapuscinski, A. R., and E. M. Hallerman. 1991. Implications of introduction of transgenic fish into natural ecosystems. Canadian Journal Fisheries and Aquatic Science 48:99–107.

Kapuscinski, A. R., T. Nega, and E. M. Hallerman. 1999. Adaptive biosafety assessment and management regimes for aquatic genetically modified organisms in the environment. Pp. 225–251 in Towards Policies for Conservation and Sustainable Use of Aquatic Genetic Resources, Conference Proceedings 59, R. S. V. Pullin, D. M. Bartley, and J. Kooiman, eds. Makati City, Philippines: ICLARM (International Center for Living Aquatic Resources Management).

Kasinathan, P., J. Knott, P. Moreira, A. Burnside, D. Jerry, and J. Robl. 2001. Effect of fibroblast donor cell age and cell cycle on development of bovine nuclear transfer embryos in vitro. Biology of Reproduction 64:1487–1493.

Kato,Y., T. Tani, Y. Sotomaru, K. Kurokawa, J. Kato, H. Doguchi, H. Yasue, and Y. Tsunoda. 1998. Eight calves cloned from somatic cells of a single adult. Science 282:2095–2098.

Kaufman, J. F., and S. J. Lamont. 1996. The chicken major histocompatibility complex. Pp. 35–64 in The Major Histocompatibility Complex in Domestic Animal Species, S. J. Lamont and L. B. Schook, eds. Boca Raton, FL: CRC Press.

Kempthorne O., and E. Pollak. 1970. Concepts of fitness in Mendelian populations. Genetics 64:125–145.

Kerr, D. E., K. Plaut, A. J. Bramley, C. M. Williamson, A. J. Lax, K. Moore, K. D. Wells, and R. J. Wall. 2001. Lysostaphin expression in mammary glands confers protection against staphylococcal infection in transgenic mice. Nature Biotechnology 19:66–70.

Kestin, S. C., T. G. Knowles, A. E. Tinch, and N. G. Gregory. 1992. Prevalence of leg weakness in broiler chickens and its relationship with genotype. Veterinary Record 131:190–194.

Kidwell, M. G. 1993. Lateral transfer in natural populations of eukaryotes. Annual Review of Genetics 27:235–056.

King, C. M., J. G. Innes, M. Flux, and M. O. Kimberley 1996. Population biology of small mammals in Pureora Forest Park: The feral house mouse (Mus musculus). New Zealand Journal Of Ecology 20:253–269.

Kjaer, J. B., and J. A. Mench. In press. Behavior problems associated with selection for increased production. In Poultry Breeding and Biotechnology, W. Muir and S. Aggrey, eds. Wallingford, United Kingdom: CABI International.

Klinkenborg, V. 2001. Cow parts. Discover 22(8):57–63.

Kolpin, D. W., E. T. Furlong, M. T. Meyer, E. M. Thurman, S. D. Zaugg, L. B. Barber, and H. T. Buxton. 2002. Pharmaceuticals, hormones, and other organic wastewater contaminants in U.S. streams, 1999–2000—A national reconnaissance. Environmental Science and Technology 36(6): 1202–1211.

Kono, T. 1997. Nuclear transfer and reprogramming. Reviews of Reproduction 2(2):74–80.

Kordis, D., and F. Gubensek. 1998. Unusual horizontal transfer of a long interspersed nuclear element between distant vertebrate classes. Proceedings of the National Academy of Sciences of the United States of America 95:10704–10709.

Kordis, D., and F. Gubensek. 1999. Horizontal transfer of non-LTR retrotransposons in vertebrates. Genetica 107:121–128.

Krebs, C. J., A. J. Kenney, and G. R. Singleton 1995. Movements of feral house mice in agricultural landscapes. Australian Journal of Zoology 43:293–302

Kruip, T. A. M., and J. H. G. den Dass. 1997. In vitro produced and cloned embryos: effects on pregnancy, parturition, and offspring. Theriogenology 47:43–52.

Kruip, T. A. M., M. M. Bevers, and B. Kemp. 2000. Environment of oocyte and embryo determines health of IVP offspring. Theriogenology 53:611–618.

Kuhholzer, B., and R. S. Prather. 2000. Advances in livestock nuclear transfer. Proceedings of the Society for Experimental Biology and Medicine 224:240–245.

Kuhholzer, B., R. J. Hawley, L. Lai, D. Kolber-Simonds, and R. S. Prather. 2001. Clonal lines of transgenic fibroblast cells derived from the same fetus result in different development when used for nuclear transfer in pigs. Biology of Reproduction 64(6):1695–1698.

Kunik, T., T. Tzfira, Y. Kapulnik, Y. Gafni, C. Dingwall, and V. Citovsky. 2001. Genetic transformation of HeLa cells by agrobacterium. Proceedings of the National Academy of Sciences of the United States of America 98(4):1871–1876.

Lacy, M. P. 2000. Broiler production-past, present, and future. Poultry Digest 59:24–26.

Lai, L., D. Kolber-Simonds, K. W. Park, H. T. Cheong, J. L. Greenstein, G. S. Im, M. Samuel, A. Bonk, A. Rieke, B. N. Day, C. N. Murphy, D. B Carter, R. J. Hawley, and R. S. Prather. 2002. Production of α-1, 3-galactosyltransferase-knockout inbred miniature swine by nuclear transfer cloning. Science 295(5557):1089–1092.

Lammers, B. P., A. J. Heinrichs, and R. S. Kensinger. 1999. The effects of accelerated growth rates and estrogen implants in prepubertal Holstein heifers on growth, feed efficiency, and blood parameters. Journal of Dairy Science 82:1746–1752.

Lande, R. 1983. The response to selection on major and minor mutations affecting a metrical trait. Heredity 50:47–65.

Lander, E. S., L. M. Linton, B. Birren, C. Nusbaum, M. C. Zody, J. Baldwin, K. Devon, K. Dewar, M. Doyle, W. FitzHugh, R. Funke, D. Gage, K. Harris, A. Heaford, J. Howland, L. Kann, J. Lehoczky, R. LeVine, P. McEwan, K. McKernan, J. Meldrim, J. P. Mesirov, C. Miranda, W. Morris, J. Naylor, C. Raymond, M. Rosetti, R. Santos, A. Sheridan, C. Sougnez, N. Stange-Thomann, N. Stojanovic, A. Subramanian, D. Wyman, J. Rogers, J. Sulston, R. Ainscough, S. Beck, D. Bentley, J. Burton, C. Clee, N. Carter, A. Coulson, R. Deadman, P. Deloukas, A. Dunham, I. Dunham, R. Durbin, L. French, D. Grafham, S. Gregory, T. Hubbard, S. Humphray, A. Hunt, M. Jones, C. Lloyd, A. McMurray, L. Matthews, S. Mercer, S. Milne, J. C. Mullikin, A. Mungall, R. Plumb, M. Ross, R. Shownkeen, S. Sims, R. H. Waterston, R. K. Wilson, L. W. Hillier, J. D. McPherson, M. A. Marra, E. R. Mardis, L. A. Fulton, A. T. Chinwalla, K. H. Pepin, W. R. Gish, S. L. Chissoe, M.

C. Wendl, K. D. Delehaunty, T. L. Miner, A. Delehaunty, J. B. Kramer, L. L. Cook, R. S. Fulton, D. L. Johnson, P. J. Minx, S. W. Clifton, T. Hawkins, E. Branscomb, P. Predki, P. Richardson, S. Wenning, T. Slezak, N. Doggett, J. F. Cheng, A. Olsen, S. Lucas, C. Elkin, E. Uberbacher, M. Frazier, R. A. Gibbs, D. M. Muzny, S. E. Scherer, J. B. Bouck, E. J. Sodergren, K. C. Worley, C. M. Rives, J. H. Gorrell, M. L. Metzker, S. L. Naylor, R. S. Kucherlapati, D. L. Nelson, G. M. Weinstock, Y. Sakaki, A. Fujiyama, M. Hattori, T. Yada, A. Toyoda, T. Itoh, C. Kawagoe, H. Watanabe, Y. Totoki, T. Taylor, J. Weissenbach, R. Heilig, W. Saurin, F. Artiguenave, P. Brottier, T. Bruls, E. Pelletier, C. Robert, P. Wincker, D. R. Smith, L. Doucette-Stamm, M. Rubenfield, K. Weinstock, H. M. Lee, J. Dubois, A. Rosenthal, M. Platzer, G. Nyakatura, S. Taudien, A. Rump, H. Yang, J. Yu, J. Wang, G. Huang, J. Gu, L. Hood, L. Rowen, A. Madan, S. Qin, R. W. Davis, N. A. Federspiel, A. P. Abola, M. J. Proctor, R. M. Myers, J. Schmutz, M. Dickson, J. Grimwood, D. R. Cox, M. V. Olson, R. Kaul, N. Shimizu, K. Kawasaki, S. Minoshima, G. A. Evans, M. Athanasiou, R. Schultz, B. A. Roe, F. Chen, H. Pan, J. Ramser, H. Lehrach, R. Reinhardt, W. R. McCombie, M. de la Bastide, N. Dedhia, H. Blocker, K. Hornischer, G. Nordsiek, R. Agarwala, L. Aravind, J. A. Bailey, A. Bateman, S. Batzoglou, E. Birney, P. Bork, D. G. Brown, C. B. Burge, L. Cerutti, H. C. Chen, D. Church, M. Clamp, R. R. Copley, T. Doerks, S. R. Eddy, E. E. Eichler, T. S. Furey, J. Galagan, J. G. Gilbert, C. Harmon, Y. Hayashizaki, D. Haussler, H. Hermjakob, K. Hokamp, W. Jang, L. S. Johnson, T. A. Jones, S. Kasif, A. Kaspryzk, S. Kennedy, W. J. Kent, P. Kitts, E. V. Koonin, I. Korf, D. Kulp, D. Lancet, T. M. Lowe, A. McLysaght, T. Mikkelsen, J. V. Moran, N. Mulder, V. J. Pollara, C. P. Ponting, G. Schuler, J. Schultz, G. Slater, A. F. Smit, E. Stupka, J. Szustakowski, D. Thierry-Mieg, J. Thierry-Mieg, L. Wagner, J. Wallis, R. Wheeler, A. Williams, Y. I. Wolf, K. H. Wolfe, S. P. Yang, R. F. Yeh, F. Collins, M. S. Guyer, J. Peterson, A. Felsenfeld, K. A. Wetterstrand, A. Patrinos, M. J. Morgan, and J. Szustakowki. 2001. Initial sequencing and analysis of the human genome. Nature 409(6822):860–921.

Lange, I. G., A. Daxenberger, and H. H. Meyer. 2001. Hormone contents in peripheral tissues after correct and off-label use of growth promoting hormones in cattle: Effect of the implant preparations Filaplix-H, Raglo, and Synovex Plus. APMIS (Acta Pathologica Microbiologica et Immunologica Scandinavica) 109:53–65.

Lanza, R. P., J. B. Cibelli, C. Blackwell, V. J. Cristofalo, M. K. Franis, G. M. Baerlocher, J. Mak, M. Schertzer, E. A. Chavez, N. Sawyer, P. M. Lansdorf, and M. D. West. 2000. Extension of cell life-span and telomere length in animals cloned from senescent somatic cells. Science 288:665–669.

Lanza, R., J. B. Cibelli, D. Faber, R. W. Sweeney, B. Henderson, W. Nevala, M. D. West, and P. J. Wettstein. 2001. Cloned cattle can be healthy and normal. Science 294:1893–1894.

Lem, A. 1999. Aquaculture and trade. In Conference on Aquaculture, Economics and Marketing. Available online at *www.globefish.org/presentations/ onepagepresentations/aquaculture.htm.*

Lewin, H. A., G. C. Russell, and E. J. Glass. 1999. Comparative organization and function of the major histocompatibility complex of domesticated cattle. Immunological Reviews 167:145–158.

Lewis, D. B., H. D. Liggitt, E. L. Effmann, S. T. Motley, S. L., Teitelbaum, K. L. Jepsen, S. A. Goldstein, J. Bonadio, J. Carpenter, and R. M. Perlmutter. 1993. Osteoporosis induced in mice by overproduction of interleukin 4. Proceedings of the National Academy of Sciences of the United States of America 90:11618–11622.

Litscher, E. S., C. Liu, Y. Echelard, and P. M. Wassarman. 1999. *Zona pellucida* glycoprotein mZP3 produced in milk of transgenic mice is active as a sperm receptor, but can be lethal to newborns. Transgenic Research 8:361–369.

Liu, J., Z. Nagy, J. Joris, H. Tournaye, P. Devroey, and A. Van Steirteghem. 1995. Successful fertilization and establishment of pregnancies after intracytoplasmic sperm injection in patients with globozoospermia. Human Reproduction 10:262–629.

Lois, C., E. J. Hong, S. Pease, E. J. Brown, and D. Baltimore. 2002. Germline transmission and tissue-specific expression of transgenes delivered by lentiviral vectors. Science 295:868–872.

Lu, J. K., T. T. Chen, S. K. Allen, Jr., T. Matsubara, and J. C. Burns. 1996. Production of transgenic dwarf surfclams, *Mulinia lateralis*, with pantropic retroviral vectors. Proceedings of the National Academy of Sciences of the United States of America. 93:3482–3486.

Lu, K. H., D. G. Cran, and G. E. Seidel, Jr. 1999. In vitro fertilization with flow-cytometrically-sorted bovine sperm. Theriogenology 52:1393–1405.

Lubber, M. S. December 4, 2000. A request for formal consultation under ESA §7 regarding EPA's delegation of the NPDES program. A letter from EPA Region I Administrator Mindy S. Lubber to Dr. Mamie A. Parker, U.S. Fish and Wildlife Service, and Ms. Patricia A. Kurkul, National Marine Fisheries Service.

Lubon, H. 1998. Transgenic animal bioreactors in biotechnology and production of blood proteins. Biotechology Annual Review 4:1–54.

Lynch M., and B. Walsh. 1998. Genetics and analysis of quantitative traits. Sunderland, MA: Sinauer Associates.

Lynch, M., and M. O'Hely. 2001. Captive breeding and the genetic fitness of natural populations. Conservation Genetics 2:363–378.

Maclean, N., and D. Penman. 1990. The application of gene manipulation to aquaculture. Aquaculture 85:35–50.

Majeskie, J. L. 1996. Status of United States dairy cattle. In the National Cooperative Dairy Herd Improvement Program. Fact Sheet K-7. Washington, DC: U.S. Department of Agriculture.

Martin, M. 1991. Socioeconomic aspects of agricultural biotechnology. Phytopathology 81(3):356–360.

Martinez, S. 1999. Vertical coordination in the pork and broiler industries: Implications for pork and chicken products. Pp. 48 in Agricultural Economics Report No. 777. Available online at *http://www.ers.usda.gov/publications/AER777/*. Accessed January 21, 2002.

Mason, G., and M. Mendl. 1993. Why is there no simple way of measuring animal welfare? Animal Welfare 2:301–319.

Massoud, M., J. Attal, D. Thepot, H. Pointu, M. G. Stinnakre, and M. C. Theron. 1996. The deleterious effects of human erythropoeitin gene driven by the rabbit whey acidic protein gene promoter in transgenic rabbits. Reproduction Nutrition Development 36:555–563.

Matthews, L. R. 1992. Ethical, moral, and welfare implications of embryo manipulation technology. ACCART (Australians Council for the Care of Animals in Research and Teaching) News 5:6–7.

Mayne, C. S., and J. McEvoy. 1993. In vitro fertilized embryos: Implications for the dairy herd. Pp. 75–83 in The Veterinary Annual 33. M. E. Raw, and T. J. Parkinson, eds. London: Blackwell Scientific.

McCreath, K. J., J. Howcroft, K. H. S. Campbell, A. Colman, A. E. Schnieke, and A. J. Kind. 2000. Production of gene-targeted sheep by nuclear transfer from cultured somatic cells. Nature 405(6790):1066–1069.

McEvoy, T. G., K. D. Sinclair, P. J. Broadbent, K. L. Goodhand, and J. J. Robinson. 1998. Post-natal growth and development of Simmenthal calves derived from in vivo or in vitro embryos. Reproduction, Fertility and Development 10:459–464.

McNeish, J. D., W. J. Scott, and S. S. Potter. 1988. Legless, a novel mutation found in PHT-1 transgenic mice. Science 241:837–839.

Meisler, M. H. 1992. Insertional mutation of "classical" and novel genes in transgenic mice. Trends in Genetics 8:341–344.

Mench, J. A. 1999. Ethics, animal welfare, and transgenic farm animals. Pp. 251–268 in Transgenic Animals in Agriculture, J. D. Murray, G. B. Anderson, A. M. Oberbauer, and M. M. McGloughlin, eds. Wallingford, UK: CABI International.

Mench, J. A. 2002. Broiler breeders: Feed restriction and welfare. World's Poultry Science Journal 58:23–29.

Mench, J. A., J. Morrow-Tesch, and L. Chu. 1998. Environmental enrichment for farm animals. Lab Animal 27:3–7.

Mendieta, R., N. L. Nagy, and F. A. Lints 1997. The potential allergenicity of novel foods. Journal of the Science of Food and Agriculture 75(4):405–411.

Meyerholz, G. W. 1983. Agriculture programs, livestock and veterinary sciences. In Residue Avoidance Program, summary. Washington, D.C.: U.S. Department of Agriculture Extension Service, Office of the Administrator.

MGD (Mouse Genome Database). 2002. In Mouse Genome Informatics Glossary. Bar Harbor, Maine: The Jackson Laboratory. Available online at www.informatics.jax.org/mgihome/other/glossary.html. Accessed June 2002.

Middendorf, G., M. Skladany, E. Ransom, and L. Busch. 2000. New agricultural biotechnologies: The struggle for democratic choice. Pp. 107–123 in Hungry for Profit: The Agribusiness Threat to Farmers, Food, and the Environment, F. Magdoff, J. B. Foster, and F. H. Buttel, eds. New York: Monthly Review Press.

Miller, A. D. 1997. Development and application of retroviral vectors. Pp. 437–473 in Retroviruses, J. M. Coffin, S. H. Hughes, and H. E. Varmus, eds. Cold Spring Harbor, NY: Cold Spring Harbor Laboratory Press.

Mittelmark, J., and A. Kapuscinski. 2001. Induced Reproduction in fish. In Minnesota Sea Grant Outreach, Aquaculture. Available online at www.seagrant.umn.edu/aqua/induce.html.

Molbak, K., D. L. Baggesen, F. M. Aarestrup, J. M. Ebbesen, J. Engberg, K. Frydendahl, P. Gerner-Smidt, A. M. Petersen, and H. C. Wegener. 1999. An outbreak of multidrug-resistant, ouinolone-resistant salmonella enterica serotype typhimurium Dt104. New England Journal of Medicine 314(19):1420–1425.

Moore, C. J., and T. B. Mepham. 1995. Transgenesis and animal welfare. ATLA (Alternatives to Laboratory Animals) 23:380–397.

Mori, T., and R. H. Devlin. 1999. Transgene and host GH gene expression in pituitary and nonpituitary tissues of normal and GH transgenic salmon. Molecular and Cellular Endocrinology 149:129–139.

Morton, D., R. James, and J. Roberts. 1993. Issues arising from recent advances in biotechnology: Report of the British Veterinary Association Foundation study group. Veterinary Record 17:53–56.

Muir, W. M., and R. D. Howard. 1999. Possible ecological risks of transgenic organism release when transgenes affect mating success: Sexual selection and the Trojan gene hypothesis. Proceedings of the National Academy of Sciences of the United States of America 24:13853–13856.

Muir, W. M., and R. D. Howard. 2001. Fitness components and ecological risk of transgenic release: A model using Japanese medaka (Oryzias latipes). American Naturalist 158:1–16.

Muir, W. M., and R. D. Howard. 2002a. Methods to assess ecological risks of transgenic fish releases. Pp. 355–383 in Genetically Engineered Organisms: Assessing Environmental and Human Health Effects, D.K. Letourneau and B. E. Burrows, eds. Boca Raton, FL: CRC Press.

Muir, W. M., and R. D. Howard. 2002b. Assessment of possible ecological risks and hazards of transgenic fish with implications for other sexually reproducing organisms. Transgenic Research 11(2):101–114..

Muller, K., H. Heller, and W. Doerfler. 2001. Foreign DNA integration: Genome-wide perturbations of methylation and transcription in the recipient genomes. Journal of Biological Chemistry 276(17):14271–14278.

Müller, M., and G. Brem. 1994. Transgenic strategies to increase disease resistance in livestock. Reproduction, Fertility, and Development 6:605–613.

Murray, J. D., and E. A. Maga. 1999. Changing the composition and properties of milk. Pp. 193–208 in Transgenic Animals in Agriculture, J. D. Murray, G. B. Anderson, A. M. Oberbauer, and M. M. McGloughlin, eds. Wallingford, UK: CABI International.

NAAB (National Association of Animal Breeders). 1996. Breakthroughs in biotechnology: research equips producers with an array of genetic improvement tools. In Electronic Resource Guide. Available online at *www.naab-css.org/education/biotech/html*.

Nancarrow, C. D., J. T. A. Marshall, J. L. Clarkson, J. D. Murray, R. M. Millard, C. M. Shanahan, P. C. Wynn, and K. A. Ward. 1991. Expression and physiology of performance regulating genes in transgenic sheep. Journal of Reproduction and Fertility 43:277S–291S.

Nebel, R. L., and S. M. Jobst. 1998. Evaluation of systematic breeding programs for lactating dairy cows: A review. Journal of Dairy Science 68:1585–1592.

Nelson, R. J. 1997. The use of genetic "knockout" mice in behavioral endocrinology research. Hormones and Behavior 31:188–196.

Niemann, H., and W. A. Kues. 2000. Transgenic livestock: Premises and promises. Animal and Reproduction Science 60–61:277–293.

Niemann, H., R. Halter, J. W. Carnwath, D. Herrmann, E. Lemme, and P. Dieter. 1999. Expression of human blood clotting factor VIII in the mammary gland of transgenic sheep. Transgenic Research 8:237–247.

Norman, H. D., T. J. Lawlor, and J. R. Wright. 2002. Performance of Holstein clones in the United States. World Congress of Animal Production. Submitted for presentation.

North, M. D., and D. B. Bell. 1990. Commercial Chicken Production Manual. New York: Van Nostrand Reinhold.

Notter, D. R. 1999. The importance of genetic diversity in livestock populations of the future. Journal of Animal Science 77:61–69.

Nottle, M. B., H. Nagashima, P. J. Verma, Z. T. Du, C. G. Grup, S. M. McIlfatrick, R. J. Ashman, M. P. Harding, C. Giannakis, P. L. Wigley, I. G. Lyons, D. T. Harrison, B. G. Luxford, R. G. Campbell, R. J. Crawford, and A. J. Robins. 1999. Pp. 145–156 in Transgenic Animals in Agriculture, J. D. Murray, G. B. Anderson, A. M. Oberbauer, and M. M. McGloughlin, eds. Wallingford, UK: CABI International.

NRC (National Research Council). 1983. Risk Assessment in the Federal Government: Managing the Process. Washington, DC: National Academy Press.

NRC (National Research Council). 1989. Field Testing Genetically Modified Organisms: Framework for Decisions. Washington, DC: National Academy Press.

NRC (National Research Council). 1996. Understanding Risk: Informing Decisions in a Democratic Society. Washington, DC: National Academy Press.

NRC (National Research Council). 1999. The Use of Drugs in Food Animals: Benefits and Risks. Washington, DC: National Academy Press.

NRC (National Research Council). 2000. Genetically Modified Pest-Protected Plants: Science and Regulation. Washington, DC: National Academy Press.

NRC (National Research Council). 2002a. Environmental Effects of Transgenic Plants: The Scope and Adequacy of Regulation. Washington, DC: National Academy Press.

NRC (National Research Council). 2002b. Stem Cells and the Future of Regenerative Medicine. Washington, DC: National Academy Press.

Onions, D., D. K. Cooper, T. J. Alexander, C. Brown, E. Claassen, J. E. Foweraker, D. L. Harris, B. W. Mahy, P. D. Minor, A. D. Osterhaus, P. P. Pastoret, and K. Yamanouchi. 2000. An approach to the control of disease transmission in pig-to-human xenotransplantation. Xenotransplantation 7(2):143–155.

Orlans, F. B. 2000. Research on animals, law, legislative, and welfare issues in the use of animals for genetic engineering and xenotransplantation. In Encyclopedia of Ethical, Legal, and Policy Issues in Biotechnology, T. H. Murray and M. J. Mehlan, eds. New York: John Wiley & Sons.

OSTP (Office of Science and Technology Policy). 2001. Case study one: Growth-enhanced salmon. Available online at http://www.ostp.gov/html/OSTP_docarchives.html.

Paine, R. T. 1966. Food web complexity and species diversity. The American Naturalist 100:65–75.

Pannell, D., and J. Ellis. 2001. Silencing of gene expression: implications for design of retrovirus vectors. Reviews in Medical Virology 11(4):205–217.

Paradis, K., G. Langford, Z. Long, W. Heneine, P. Sandstrom, W. M. Switzer, L. E. Chapman, C. Lockey, D. Onions, and E. Otto. 1999. Search for cross-species transmission of porcine endogenous retrovirus in patients treated with living pig tissue: The XEN 111 study group. Science 285(5431):1236–1241.

Parks, K. R., E. J. Eisen, and J. D. Murray. 2000a. Correlated responses to selection for large body size in oMt1a-oGH mice: Growth, feed efficiency, and body composition. Journal of Animal Breeding Genetics 117:385–405.

Parks, K. R., E. J. Eisen, I. J. Parker, L. G. Hester, and J. D. Murray. 2000b. Correlated responses to selection for large body size in oMt1a-oGH transgenic mice: Reproductive traits. Genetical Research (Cambridge) 75:199–208.

Payne, L. N., A. M. Gillespie, and K. Howes. 1992. Myeloid leukaemogenicity and transmission of the HPRS-103 strain of avian leukosis virus. Leukemia 6(11):1167–1176.

Perry, A. C., T. Wakayama, H. Kishikawa, T. Kasia, M. Okabe, Y. Toyoda, and R. Yanagimiachi. 1999. Mammalian transgenesis by intracytoplasmic sperm injection. Science 284:1180–1183.

Petters, R. M., C. A. Alexander, K. D. Wells, E. B. Colling, J. R. Sommer, M. R. Blanton, G. Rojas, H. Y. Flowers, W. L. Banin, E. Cideciyan, A. V. Jacobson, and S. G. Wong. 1997. Genetically engineered large animal model for studying cone photoreceptor survival and degeneration in retinitis pigmentosa. Nature Biotechnology 15:965–970.

Pfeifer, A., M. Ikawa, Y. Dayn, and I. M. Verma. 2002. Transgenesis by lentiviral vectors: Lack of gene silencing in mammalian embryonic stem cells and preimplantation embryos. Proceedings of the National Academy of Sciences of the United States of America 99:2140–2145.

Pimm, S. L. 1984. The complexity and stability of ecosystems. Nature 307:321–326.

Pinstrup-Andersen, P., and R. Pandya-Lorch. 1999. Securing and sustaining adequate world food production for the third millennium. Pp. 27–48 in NABC Report 11, World Food Security and Sustainability: The Impacts of Biotechnology and Industrial Consolidation, D. P. Weeks, J. B. Segelken, and R. W. F. Hardy, eds. Ithaca, New York: National Agricultural Biotechnology Council.

Pisenti, J. M., M. E. Delany, R. L. Taylor, Jr., U. K. Abbott, H. Abplanalp, J. A. Arthur, M. R. Bakst, C. Baxter-Jones, J. J. Bitgood, F. A. Bradley, K. M. Cheng, R. R. Dietert, J. B. Dodgson, A. M. Donoghue, A. B. Emsley, R. J. Etches, R. R. Frahm, R. J. Gerrits, P. F. Goetinck, A. A. Grunder, D. E. Harry, S. J. Lamont, G. R. Martin, P. E. McGuire, G. P. Moberg, L. J. Pierro, C. O. Qualset, M. A. Qureshi, F. T. Shultz, and B. W. Wilson. 1999. Avian genetic resources at risk: An assessment and proposal for conservation of genetic stocks in the USA and Canada. Report No. 20. Davis, CA: University of California Division of Agriculture and Natural Resources, Genetic Resources Conservation Program.

Polejaeva, I. A., S. H. Chen, T. D. Vaught, R. L. Page, J. Mullins, S. Ball, Y. Dai, J. Boone, S. Walker, D. L. Ayares, A. Colman, and K. H. Campbell. 2000. Cloned pigs produced by nuclear transfer from adult somatic cells. Nature 407(6800):86–90.

PPL Therapeutics. 2001. Interim Report. Blacksburg, VA: PPL Therapeutics, PLC.

Prather, R. S., F. L. Barnes, M. M. Sims, J. M. Robl, W. H. Eyestone, and N. L. First. 1987. Nuclear transplantation in the bovine embryo: assessment of donor nuclei and recipient oocyte. Biology of Reproduction 37:859–866.

Prather, R. S., T. Tao, and Z. Machaty. 1999. Development of the techniques for nuclear transfer in pigs. Theriogenology 51(2):487–498.

Pryce, J. E., M. P. Coffey, and S. Brotherstone. 2000. The genetic relationship between calving interval, body condition, score, and linear type and management traits in registered Holsteins. Journal of Dairy Science 83:2664–2671.

Purcell, D. F., C. M. Broscius, E. F. Vanin, C. E. Buckler, A. W. Nienhuis, and M. A. Martin. 1996. An array of murine leukemia virus-related elements is transmitted and expressed in a primate recipient of retroviral gene transfer. Journal of Virology 70:887–897.

Pursel, V. G., C. E. Rexroad, J. Bolt, K. F. Miller, R. J. Wall, R. E. Hammer, K. A. Pinkert, R. D. Palmiter, and R. L. Brinster. 1987. Progress on gene transfer in farm animals. Veterinary Immunology and Immunopathology 17:303–312.

Pursel, V. G., D. J. Bolt, K. F. Miller, C. A. Pinkert, R. E. Hammer, R. D. Palmiter, and R. L. Brinster. 1990. Expression and performance in transgenic pigs. Journal of Reproduction and Fertility 40:235S–245S.

Pursel, V. G., K. A. Pinkert, K. F Miller, D. J. Bolt, R. G. Campbell, R. D. Palmiter, R. L. Brinster, and R. E. Hammer. 1989. Genetic engineering of livestock. Science 244:1281–1288.

Pursel, V. G., P. Sutrave, R. J. Wall, A. M. Kelly, and S. H. Hughes. 1992. Transfer of c-ski gene into swine to enhance muscle development. Theriogenology 37:278.

Racaniello, V. R., and R. Ren. 1994. Transgenic mice and the pathogenesis of poliomyelitis. Archives of Virology 9:79S–86S.

Rahman, M. A., and N. Maclean. 1999. Growth performance of transgenic tilapia containing an exogenous piscine growth hormone gene. Aquaculture 173:333–346.

Reddy, V. B., J. Vitale, C. Wei, M. Montoya-Zavala, S. L. Stice, and J. Balise. 1991. Expression of human growth hormone in the milk of transgenic mice. Animal Biotechnology 2:5–29.

Regal, P. J. 1986. Models of genetically engineered organisms and their ecological impact. Pp. 111–129 in Ecology of Biological Invasions of North America and Hawaii, H. A. Mooney and J. A. Drake, eds. New York: Spring-Verlag.

Reik, W., I. Römer, S. C. Barton, M. A. Surani, S. K. Howlett, and J. Klose. 1993. Adult phenotype in the mouse can be affected by epigenetic events in the early embryo. Development 199:933–942.

Renard, J. P., S. Chastant, P. Chesné, C. Richerd, J. Marchal, N. Cordonnier, P. Chavatte, and X. Vignon. 1999. Lymphoid hypoplasia and somatic cloning. The Lancet 353:1489–1491.

Rexroad, C. E. 1994 Transgenic farm animals. ILAR (Institute for Laboratory Animal Research) Journal 36:5–9.

Ricklefs, R. E. 1990. Ecology, Third Edition. New York: W. H. Freeman.

Rideout, W. M., K. Eggan, and R. Jaenisch. 2001. Nuclear cloning and epigenetic reprogramming of the genome. Science 293(5532):1093–1098.

Robertson, H. M. 1997. Multiple *mariner* transposons in flatworms and hydras are related to those of insects. Journal of Heredity 88:195–201.

Robertson, H. M., and D. J. Lampe. 1995. Recent horizontal transfer of a mariner transposable element among and between Diptera and Neuroptera. Molecular Biology and Evolution 12:850–862.

Robertson, H. M., and K. L. Zumpano. 1997. Molecular evolution of an ancient mariner transposon, *Hsmar1*, in the human genome. Gene 205:203–217.

Rollin, B. E. 1995. The Frankenstein Syndrome: Ethical and Social Issues in the Genetic Engineering of Animals. Cambridge: Cambridge University Press.

Rollin, B. E. 1986. The frankenstein thing: The moral impact of genetic engineering of agricultural animals on society and future science. Basic Life Science 37:285–297.

Römer, I., W. Reik, W. Dean, and J. Klose. 1997. Epigenetic inheritance in the mouse. Current Biology 7:277–280.

Rosenberg, N., and P. Jolicoeur. 1997. Retroviral pathogenesis. Pp. 475–585 in Retroviruses, J. M. Coffin, S. H. Hughes, and H. E. Varmus. Cold Spring Harbor, NY: Cold Spring Harbor Laboratory Press.

Royal, M., G. E. Mann, and A. P. Flint. 2000. Strategies for reversing the trend towards subfertility in dairy cattle. Veterinary Journal 160:53–60.

Saif, L. J., and M. B. Wheeler. 1998. WAPping gastroenteritis with transgenic antibodies. Nature Biotechnolgy 16:334–335.

Salminen, S., C. Bouley, M. C. Boutron-Ruault, J. H. Cummings, A. Frank, G. F. R. Gibson, E. Isolauri, and M. C. Moreau. 1998. Functional food science and gastrointestinal physiology and function. British Journal of Nutrition 80:147S–171S.

Sampson, H. A. 1997. Food Allergy. Journal of the American Medical Association 278(22):1888–1894.

Sandler, L., and E. Novitski. 1957. Meiotic drive as an evolutionary force. American Naturalist 91:105.

Sandrin, M. S., H. A. Vaughan, P. L. Dabkowski, and I. F. C. McKenzie. 1993. Anti-pig antibodies in human serum react predominantly with Galα (1,3) Gal epitopes. Proceedings of the National Academy of Science of the United States of America 90:11391–11395.

Sarmasik, A., C. Z. Chun, I. K. Jang, J. K. Lu, and T. T. Chen. 2001. Production of transgenic live-bearing fish and crustaceans with replication-defective pantropic retroviral vectors. Marine Biotechnology 3:S177–S184.

Sarmasik A., G. Warr, and T. T. Chen. 2002. Production of transgenic medaka with increased resistance to bacterial pathogens. Marine Biotechnology 4(3):310–322.

Saunders, R. L., G. L. Fletcher, and C. L. Hew. 1998. Smolt development in growth hormone transgenic Atlantic salmon. Aquaculture 168:177–193.

Scadden, D. T., B. Fuller, and J. M. Cunningham. 1990. Human cells infected with retrovirus vectors acquire an endogenous murine provirus. Journal of Virology 64:424–427.

Scarpa, J., J. E. Toro, and K. T. Wada. 1994. Direct comparison of six methods to induce triploidy in bivalves. Aquaculture 119:119–133.

Schiffer, B., A. Daxenberger, K. Meyer, and H. Meyer. 2001. The fate of trenbolone acetate and melengestrol acetate after application of growth promoters in cattle: Environmental studies. Environmental Health Perspectives 109:1145–1151.

Schmidt, M., T. Greve, B. Avery, J. F. Beckers, J. Sulon, and B. Hansen. 1996. Pregnancies, calves, and calf viability after transfer of in vitro produced bovine embryos. Theriogenology 46:527–539.

Schook, L. B., M. S. Rutherford, J. K. Lee, Y. C. Shia, M. Bradshaw, and J. K. Lunney. 1996. The swine major histocompatibility complex. Pp. 213–244 in The Major Histocompatibility Complex Region of Domestic Animal Species, L. B. Schook and S. J. Lamont, eds. Boca Raton, FL: CRC Press.

Schroder, S. L. 1982. The Role of Sexual Selection in Determining Overall Mating Patterns and Mate Choice in Chum Salmon. Ph.D. dissertation. Seattle: University of Washington.

Schwartz, F. J. 1972. World literature to fish hybrids with an analysis by family, species and hybrid: Supplement 1. National Oceanic and Atmospheric Administration Technical Report NMFS SSRF-750. Washington, DC: National Marine Fisheries Service.

Scientists' Working Group on Biosafety. 1998. Manual for Assessing Ecological and Human Health Effects of Genetically Engineered Organisms. Part One: Introductory Text and Flowcharts. Part Two: Flowcharts and Worksheets. Edmonds Institute 245. Available online at www.edmonds-institute.org/manual.html.

Seamark, R. F. 1993. Recent advances in animal biotechnology: Welfare and ethical implications. Livestock Production Science 36:5–15.

Seidel, G. E., Jr. 1984. Applications of embryo transfer and related technologies to cattle. Journal of Dairy Science 67:2786–2796.

Seidel, G. E, Jr. 2001. Cloning, transgenesis, and genetic variance in animals. Cloning and Stem Cells 4:251–256.

Shamay, A., V. G. Pursel, E. Wilkinson, R. J. Wall, and L. Hennighausen. 1992. Expression of the whey acidic protein in transgenic pigs impairs mammary development. Transgenic Research 1:124–132.

Sharma, A., M. J. Martin, J. F. Okabe, R. A. Truglio, N. K. Dhan, J. S. Logan, and R. Khumar. 1994. An isologous porcine promoter permits high level expression of human hemoglobin in transgenic swine. Biotechnology 12:55–59.

Sharma, V. 1999. Current scenario of malaria in India. Parassitologia 41(1/3):349–353.

Sherman, A., A. Dawson, C. Mather, H. Gilhooley, Y. Li, R. Mitchell, D. Finnegan, and H. Sang. 1998. Transposition of the Drosophila element mariner into the chicken germline. Nature Biotechnology 16:1050–1053.

Shiels, P. G., A. J. Kind, K. H. S. Campbell, D. Waddington, I. Wilmut, A. Coleman, and A. E. Schnieke. 1999. Analysis of telomere lengths in cloned sheep. Nature 399:316–317.

Siewerdt, F., E. J. Eisen, and J. D. Murray. 2000a. Correlated changes in fertility and fitness traits in lines of oMt1a-oGH transgenic mice selected for increased 8-week body weight. Journal of Animal Breeding and Genetics 117:83–95.

Siewerdt, F., E. J. Eisen, J. D. Murray, and I. J. Parker. 2000b. Response to 13 generations of selection for increased 8-week body weight in lines of mice carrying a sheep growth hormone-based transgene. Journal of Animal Science 78:832–845.

Simon, R. A. 1985. Allergic and other adverse reactions to foods. Gastroenterology. 7:4435–4449.

Sinclair, K. D., L. E. Young, I. Wilmut, and T. G. McEvoy. 2000. In-utero overgrowth in ruminants following embryo culture: Lessons from mice and a warning to men. Human Reproduction 15:68S–86S.

Sinclair, K. D., T. G. McEvoy, E. K. Maxfield, C. A. Maltin, L. E. Young, I. Wilmut, P. J. Broadbent, and J. J. Robinson. 1999. Aberrant fetal growth and development after in vitro culture of sheep zygotes. Journal of Reproduction and Fertility 116:177–186.

Smith, A. G. 2001. Embryo-derived stem cells: Of mice and men. Annual Review of Cell and Developmental Biology 17:435–462.

Sola, I., J. Castilla, B. Pintado, J. M. Sanchez-Morgado, C. B. A. Whitelaw, A. J. Clark, and L. Enjuanes. 1998. Transgenic mice secreting coronavirus neutralizing antibodies into the milk. Journal of Virology 72:3762–3772.

Soriano, P., R. D. Cone, R. C. Mulligan, and R. Jaenisch. 1986. Tissue-specific and ecotropic expression of genes introduced into transgenic mice by retroviruses. Science 234:1409–1413.

Sparks, P., R. Shepherd, and L. Frewer. 1995. Assessing and structuring attitudes towards the use of gene technology in food production: The role of perceived ethical obligation. Journal of Basic and Applied Social Psychology 16:267–285.

Spielman, A. 1994. Why entomological anti-malarial research should not focus on transgenic mosquitos. Parasitology Today 10:374.

Spielman, A., J. C. Beier, and A. E. Kiszewski. 2002. Ecological and community considerations in engineering arthropods to suppress vector-borne disease. Pp. 315–329 in Genetically Engineered Organisms: Assessing Environmental and Human Health Effects, D. K. Letourneau and B. E. Burrows, eds. Boca Raton, FL: CRC Press.

Stephenson, J. 2002. Researchers wrestle with spread and control of emerging infections. Journal of the American Medical Association 287(16):2061–2064.

Stevens, E. D., A. Sutterlin, and T. Cook. 1998. Respiratory metabolism and swimming performance in growth hormone transgenic Atlantic salmon. Canadian Journal of Fisheries and Aquatic Sciences 55:2028–2035.

Stevens, E. D., and A. Sutterlin. 1999. Gill morphology in growth hormone transgenic salmon. Environmental Biology of Fishes 54:415–411.

Stice, S. L., J. M. Robl, F. A. Ponce de Leon, J. Jerry, P. G. Golueke, J. B. Cibelli, and J. J. Kane. 1998. Cloning: New breakthroughs leading to commercial opportunities. Theriogenology 49(1):129–138.

Stowers, A. W., L. Chen, Y. Zhang, M. C. Kennedy, L. Zou, T. J. Rice, D. C. Kaslow, A. Saul, C. A. Long, H. Meade, and L. H. Miller. 2001. A recombinant vaccine expressed in the milk of transgenic mice protects *Aotus* monkeys from a lethal challenge with *Plasmodium falciparum*. Proceedings of the National Academy of Sciences of the United States of America 99(1):339–344.

Stoye, J. P., C. Moroni, and J. Coffin. 1991. Virological events leading to spontaneous AKR thymomas. Journal of Virology 65:1273–1285.

Sundararajan, P., P. W. Atkinson, and D. A. O'Brochta. 1999. Transposable element interactions in insects: Crossmobilization of *hobo* and *Hermes*. Insect Molecular Biology 8:359–368.

Suszkiw, J. 2001. Scientists develop first transgenic cow clone for mastitis disease resistance. In ARS News and Information. Available online at *http://www.ars.usda.gov/is/pr/2001/010110.2.htm.*

Takeda, K. S., S. Takahashi, A. Onishi, Y. Goto, A. Miyazawa, and H. Imai. 1999. Dominant distribution of mitochondrial DNA from recipient oocytes in bovine embryos and offsring after nuclear transfer. Journal of Reproduction and Fertility 116:253–259.

Takeuchi, Y., C. Patience, S. Magre, R. A. Weiss, P. T. Banerjee, P. Le Tissier, and J. P. Stoye. 1998. Host range and interference studies of three classes of pig endogenous retrovirus. Journal of Virology 72(12):9986–9991.

Tamashiro, K. L. K., T. Wakayama, R. J. Blanchard, D. C. Blancahard, and R. Yanagimachi. 2000. Postnatal growth and behavioral development of mice cloned from adult cumulus cells. Biology of Reproduction 63:328–334.

Tauxe, R. V. 1997. Emerging foodborne diseases: An evolving public health challenge. Emerging Infectious Diseases 3(4):425–434.

Taylor, S. L., and S. L. Hefle. 2001. Will genetically modified foods be allergenic? The Journal of Allergy and Clinical Immunology 107(5):765–771.

Taylor, S. L., S. L. Hefle, and A. Munoz-Furlong. 1999. Food allergies and avoidance diets. Nutrition Today 34(1):15–22.

Tearle, R. G., M. J. Tange, Z. L. Zannettino, M. Katerelos, T. A. Shinkel, B. J. W. Van Denderen, A. J. Lonie, I. Lyons, M. B. Nottle, T. Cox, C. Becker, A. M. Peura, P. L. Wigley, R. J. Crawford, A. J. Robins, M. J. Pearse, and A. J. F. d'Apice. 1996. The α-1,3-Galactosyltransferase knockout mouse: Implications for xenotransplantation. Transplantation 61:13–19.

The Royal Society. 2001. The Use of Genetically Modified Animals. London: The Royal Society.

Thepot, D., E. Devinoy, M. L. Fontaine, M. G. Stinnakre, M. Masoud, G. Kann, and L. M. Houdebine. 1995. Rabbit whey acidic protein gene upstream region controls high-level expression of bovine growth hormone in the mammary gland of tranagenic mice. Molecular Reproduction and Development 42:261–267.

Theuring, F., M. Thunecke, U. Kosciessa, and J. D. Turner. 1997. Transgenic animals as models of neurodegenerative diseases in humans. Trends in Biotechnology 15(8):320–325.

Thompson, P. B. 2001. Food Animal Productivity and Welfare. Presentation to National Academies of Science Public Workshop, Defining Science-based Concerns Associated with Products of Animal Biotechnology. 27 Nov. Washington, DC. Available online at *http://video.nationalacademies.org/ramgen/dels/112701_d.rm*.

Thomson, J. A., J. Itskovitz-Eldor, S. S. Shapiro, M. A. Waknitz, J. J. Swiergiel, V. S. Marshall, and J. M. Jones. 1998. Embryonic stem cell lines derived from human blastocysts. Science 282(5391):1145–1147.

Thoraval, P., M. Afanassieff, F. L. Cosset, F. Lasserre, G. Verdier, F. Coudert, and G. Dambrine. 1995. Germline transmission of exogenous genes in chickens using helper-free ecotropic avian leukosis virus-based vectors. Transgenic Research 4(6):369–377.

Tiedje, J. M., R. K. Colwell, Y. L. Grossman, R. E. Hodson, R. E. Lenski, R. N. Mack, and P. J. Regal. 1989. The planned introduction of genetically engineered organisms: Ecological considerations and recommendations. Ecology 70:298–315.

Travers, A. 1999. Chromatin modification by DNA tracking. Proceedings of the National Academy of Sciences of the United States of America 96:13634–13637.

Tu, C. F., K. Tsuji, K. H. Lee, R. Chu, T. J. Sun, Y. C. Lee, C. N. Weng, and C. J. Lee. 1999. Generation of HLA-DP transgenic pigs for the study of xenotransplantation. International Surgery 84:176–182.

U.S. Public Health Service. 1978. Foodborne and Waterborne Disease Outbreaks. Washington, DC: U.S. Government Printing Office.

United States Mission to the European Union. 1999. U.S. interagency task force on beef hormones. In A Primer on Beef Hormones. Available online at *http://www.useu.be/issues/BeefPrimer022699.html*.

University of Guelph. 2001. In AnimalNet, University of Guelph Center for Food Safety. Available online at *www.plant.uoguelph.ca/safefood/archives/animalnet/2001/3-2001/an-03-08-01-01.txt*. Accessed March 8, 2002.

Urquidi, V., D. Tarin, and S. Goodison. 2000. Role of telomerase in cell senescence and oncogenesis. Annual Review of Medicine 51:65–79.

USDA. 1995. Performance standards for safely conducting research with genetically modified fish and shellfish. In Agricultural Biotechnology Research Advisory Committee (ABRAC), working group on Aquatic Biotechnology and Environmental Safety, United States Department of Agriculture. Document No. 95–01. Washington, DC: U.S. Department of Agriculture.

USDA. 2001. Agricultural Statistics. Available online at *http://www.usda.gov/nass/pubs/agstats*.

Utter, F. 2002. Genetic impacts of fish introductions. In Population Genetics: Principles and Practices for Fisheries Scientists. American Fisheries Society. Available online at *http://www.fisheries.org/publications/ /epubs/genetics/chapter15.pdf*.

Van der Laan, L. J., C. Lockey, B. C. Griffeth, F. S. Frasier, C. A. Wilson, D. E. Onions, B. J. Hering, Z. Long, E. Otto, B. E. Torbett, and D. R. Salomon. 2000. Infection by porcine endogenous retrovirus after islet xenotransplantation in SCID mice. Nature 407(6800):90–94.

Van der Lende, T., F. A. M. de Loos, and T. Jorna. 2000. Postnatal health and welfare of offspring conceived in vitro: A case for epidemiologic studies. Theriogenology 53:549–554.

Van Reenen, C. G., and H. J. Blokhuis. 1993. Investigating welfare of dairy calves involved in genetic modification: Problems and perspectives. Livestock Production Science 36:81–90.

Van Reenen, C. G., T. H. E. Meuwissen, H. Hopster, K. Oldenbroek, T. A. M. Kruip, and H. J. Blokhuis. 2001. Trangenesis may affect farm animal welfare: A case for systematic risk assessment. Journal of Animal Science 79:1763–1779.

Van Stekelenburg-Hamers, A. E. P., T. A. E. Van Achterberg, H. G. Rebel, J. E. Flechon, K. S. Campbell, S. M. Weima, and C. L. Mummery. 1995. Isolation and characterization of permanent cell lines from inner cell mass cells of bovine blastocysts. Molecular Reproduction and Development 40:444–454.

Van Wagtendonk-deLeeuw, A. M., B. J. G. Aerts, and J. H. G. den Daas. 1998. Abnormal offspring following in vitro production of bovine preimplantation embyos: A field study. Theriogenology 49:883–894.

Van Wagtendonk-deLeeuw, A. M., E. Mullaart, A. P. W. de Roos, J. S. Merton, J. H. den Daas, B. Kemp, and L. de Ruigh. 2000. Effects of different reproduction techniques: AI, MOET, or IVP, on health and welfare of bovine offspring. Theriogenology 53:575–597.

Venugopal, K. 1999. Avian leukosis virus subgroup J: A rapidly evolving group of oncogenic retroviruses. Research in Veterinary Science 67(2):113–119.

Volpe, J. P., E. B. Taylor, D. W. Rimmer, and B. W. Glickman. 2000. Evidence of natural reproduction of aquaculture-escaped Atlantic salmon in a coastal British Columbia river. Conservation Biology 14(3):899–903.

Vtorov, I. P. 1993. Feral pig removal: Effects on soil microarthropods in a Hawaiian rain forest. Journal of Wildlife Management 57:875–880.

Walker, S. K., K. M. Hartwich, and R. F. Seamark. 1996. The production of unusually large offspring following embry manipulation: concepts and challenges. Theriogenology 45:111–120.

Wall, R. J. 2001. How We Make "Them" and Where We're Headed. Presentation to National Academies of Science Public Workshop, Defining Science-based Concerns Associated with Products of Animal Biotechnology. 27 Nov. Washington, DC. Available online at *http://video.nationalacademies.org/ramgen/dels/112701_a.rm*.

Wayne, R. K., and E. A. Ostrander. 1999. Origin, genetic diversity, and genome structure of the domestic dog breeds. Bioessays 21:247–257.

Weary, D. M., M. C. Appleby, and D. Fraser. 1999. Responses of piglets to early separation from the sow. Applied Animal Behavior Science 63:289–300.

Weigel, K. A. 2001. Controlling inbreeding in modern breeding programs. Journal of Dairy Science 84:E177–E184.

Weiss, R. A. 1998. Transgenic pigs and virus adaptation. Nature 391:327–328.

Wells, D. N., P. M. Misica, and H. R. Tervit. 1999. Production of cloned calves following nucear transfer with cultured adult mural granulosa cells. Biology of Reproduction 60:996–1005.

Wells, D. N., P. M. Misica, T. A. Day, and H. R. Tervit. 1997. Production of cloned lambs from an established embryonic cell line: A comparison between in vivo- and in vitro-matured oocytes. Biology of Reproduction 57:385–393.

Westhusin, M. E., C. R. Long, T. Shin, J. R. Hill, C. R. Looney, J. H. Pryor, and J. A. Piedrahita. 2001. Cloning to reproduce desired genotypes. Theriogenology 55(1):35–49.

Wheeler, M. B., 1994. Development and validation of swine embryonic stem cells: A review. Reproduction, Fertility, and Development 6:1–6.

Whitelaw, B. 1999. Toward designer milk. Nature Biotechnology 17:135–136.

Wilkie, T. M., R. L. Brinster, and R. D. Palmiter. 1986. Germline and somatic mosaicism in transgenic mice. Developmental Biology 118:9–18.

Willadsen, S. M. 1979. A method for culture of micro manipulated sheep embryos and its use to produce monozygotic twins. Nature 277:298–300.

Willadsen, S. M. 1986. Nuclear transplantation in sheep embryos. Nature 320:63–65.

Willadsen, S. M. 1989. Cloning of sheep and cow embryos. Genome 31:956–962.

Willadsen, S. M., and C. Polge. 1981. Attempts to produce monozygotic quadruplets in cattle by blastomeres separation. Veterinary Record 108:211–213.

Williams, T. J., R. P. Elsden, and G. E. Seidel. 1984. Pregnancy rates with bisected bovine embryos. Theriogenology 22:521–531.

Wilmut, I., A. E. Schnieke, J. McWhir, A. J. Kind, and K. H. Campbell. 1997. Viable offspring derived from fetal and adult mammalian cells. Nature 385(6619):810–813.

Wilmut, I., L. Young, and K. H. Campbell. 1998. Embryonic and somatic cell cloning. Reproduction, Fertility, and Development 10(7/8):639–643.

Wilson, J. M., J. D. Williams, K. R. Bondioli, C. R. Looney, M. E. Westhuin, and D. F. McCalla. 1995. Comparison of birth weight and growth characteristics of bovine calves produced by nuclear transfer (cloning), embryo transfer and natural mating. Animal Reproduction Science 38:73–84.

Wiser, V. 1986. Healthy livestock-wholesome meat: A short history. Pp. 1–18 in One Hundred Years of Animal Health, 1884–1984, V. Wiser, L. Mark, H. Graham, D. V. Robertson, and M. M. Jacobs, eds. Journal of the National Agricultural Library Associates 11(1/4).

Wolf, E., W. Schernthaner, V. Zakhartchenko, K. Prelle, M. Stojkovic, and G. Brem. 2000. Transgenic technology in farm animals—Progress and perspectives. Experimental Physiology 85(6):615–625.

Woychik, R. P., and K. Alagraman. 1998. Insertional mutagenesis in transgenic mice generated by the pronuclear microinjection procedure. The International Journal of Developmental Biology 42(7):857–860.

Woychik, R. P., T. A. Stewart, L. G. Davis, P. D. D'Eustachio, and P. Leder. 1985. An inherited limb deformity created by insertional mutagenesis in a transgenic mouse. Nature 318:36–40.

Wright, G., A. Carver, D. Cottom, D. Reeves, A. Scott, P. Simons, I. Wilmut, I. Garner, and A. Colman. 1991. High level expression of active human alpha-1-antitrypsin in the milk of transgenic sheep. Biotechnology 9:830–834.

Wright, S. 1969. Evolution and the Genetics of Populations, Volume 2. Chicago: University of Chicago Press.

Wright, S. 1982. The shifting balance theory and macroevolution. Annual Review of Genetics 16:1–19.

Xu, A., M. J. van Eijk, C. Park, and H. A. Lewin. 1993. Polymorphism in BoLA-DRB3 exon 2 correlates with resistance to persistent lymphocytosis caused by bovine leukemia virus. Journal of Immunology 151:6977–6985.

Yoshiyma, M., Z. Tu, Y. Kainoh, H. Honda, T. Shono, and K. Kimura. 2001. Possible horizontal transfer of a transposable element from host to parasitoid. Molecular Biology and Evolution 18:1952–1958.

Young, E. 2002. GM pigs are both meat and veg. Available online at *http://www.newscientist.com/news/news.jsp?id=ns99991841*. Accessed January 25, 2002.

Young, L. E., and H. R. Fairburn. 2000. Improving the safety of embryo technologies: Possible role of genomic imprinting. Theriogenology 53:627–648.

Young, L. E., K. D. Sinclair, and I. Wilmut. 1998. Large offspring syndrome in cattle and sheep. Reviews of Reproduction 3(3):155–163.

Youngson, A. F., J. H. Webb, J. C. MacLean, and B. M. Whyte. 1997. Frequency of occurrence of reared Atlantic salmon in Scottish salmon fisheries. International Council for the Exploration of the Seas Journal of Marine Science 54:1216–1220.

Yuhki, N., and S. J. O'Brien. 1990. DNA variation of the mammalian major histocompatibility complex reflects genomic diversity and population history. Proceedings of the National. Academy of Sciences of the United States of America 87:836–840.

Zaneveld, L. J. D., D. P. Waller, R. A. Anderson, C. Chany, W. F. Rencher, K. Feathergill, X. H. Diao, G. F. Doncel, B. Herold, and M. Cooper. 2002. Efficacy and safety of a new vaginal contraceptive antimicrobial formulation containing high molecular weight poly (sodium 4-styrenesulfonate). Biology of Reproduction 66:886–894.

Zajc, I., C. S. Mellersh, and J. Sampson. 1997. Variability of canine microsatellites within and between different dog breeds. Mammalian Genome 8:182–185.

Zhou, M. F. 2002. Chromosome set instability in 1–2 year old triploid Crassostrea ariakensis in multiple environments. M. A. thesis. Gloucester Point, Virginia: Virginia Institute of Marine Science.

Zuelke, K. A. 1998. Transgenic modification of cow's milk for value-added processing. Reproduction, Fertility, and Development 10:671–676.

Glossary

Agrobacterium A pathogenic bacterium of plants that can inject a plasmid DNA (T DNA) into plant cells.

Allele One of the variant forms of a gene at a particular locus, or location, on a chromosome. Different alleles produce variation in inherited characteristics such as blood type. In an individual, the dominant form of the allele might be expressed more than the recessive one.

Allotransplantation Transplantation of cells, tissues, or organs from another member of the same species.

Biopharm animals Transgenic animals modified to produce proteins for extraction, purification, and therapeutic use.

Biopharming The production of biopharmaceuticals in domestic animals.

Biosecurity Measures to protect from infection.

Bovine spongiform encephalopathy (BSE) A disease of cattle, related to scrapie of sheep, also know as "mad cow disease." It is hypothesized to be caused by a prion, or small protein, which alters the structure of a normal brain protein, resulting in destruction of brain neural tissue.

Chimeras Animals (or embryos) composed of cells of different genetic origin.

Clone Defines both molecular clones and whole-animal clones.

Cloning The propagation of genetically exact duplicates of an organism by a means other than sexual reproduction; for example, the vegetative production of new plants or the propagation of DNA molecules by insertion into plasmids. Often, but inaccurately, used to refer to the propagation of animals by nuclear transfer.

Co-cultivation Growth of cultured cells together.

Commensal Living on or within another organism, and deriving benefit without harming the host.

Control elements DNA sequences in genes that interact with regulatory proteins (such as transcription factors) to determine the rate and timing of expression of the genes as well as the beginning and end of the transcript.

CpG methylation A heritable chemical modification of DNA (replacement of cytosine by 5-methyl cytosine) that, when present in a control region, usually suppresses expression of the corresponding gene.

CJD or Creutzfeldt-Jakob Disease A disease of humans hypothesized to be caused by a prion, or small protein, which alters the structure of a normal brain protein, resulting in destruction of brain neural tissue. The most common form is thought to have genetic origins. There is strong epidemiologic and laboratory evidence for a causal association between new variant CJD and BSE.

Cytomegalovirus A common, usually benign, herpesvirus that can cause life-threatening infection in immunosuppressed individuals.

Ecosystem disruption Any perturbation to either the structure or function of an ecosystem.

Ectopic gene expression Expression of a (trans)gene in a tissue where, or developmental stage when, such expression is not expected.

Electroporation Introduction of DNA into a cell mediated by a brief pulse of electricity.

Embryonic stem (ES) cells Cell lines derived from early embryos that have the potential to differentiate into all types of somatic cells as well as to form germ line cells, and hence whole animals, when injected into early embryos.

Endogenous provirus See **Endogenous retrovirus.**

Endogenous retrovirus Integrated retrovirus DNA (provirus) derived from infection of the germline of an ancestral animal. All animals are thought to carry numerous endogenous (but nonfunctional) retroviruses, some of which were inserted many millions of years ago.

Enucleated oocyte (cytoplast) An egg cell from which the nucleus has been removed mechanically.

Enveloped viruses Viruses whose particles (virions) are surrounded by a lipid bilayer derived by budding from the cell membrane. Examples include retroviruses, herpesviruses, influenza viruses, and many more.

Epstein-Barr virus A common and usually benign herpesvirus that is the cause of infectious mononucleosis, but can cause life-threatening infection in immunosuppressed patients.

Feral Refers to an individual or population that has returned to the wild after a history of domestication.

Fibroblast A type of relatively undifferentiated cell found in many parts of the body involved primarily in wound healing. Fibroblasts are relatively easy to grow in cell culture, and often are used for this purpose.

Fitness The ability to survive to reproductive age and produce viable offspring. Fitness also describes the frequency distribution of reproductive success for a population of sexually mature adults.

G_0 A state characterized by cells that have exited the cell cycle and entered into a resting phase.

Galactose-1,3-galactose A carbohydrate found as a modification to cell surface glycoprotein on all mammals (and many other organisms) except for old-world primates (including humans). The presence of naturally occurring antibodies to this modification in humans is a major (but not the only) cause of rejection of xenotransplanted organs.

Galactosyl transferase The enzyme, lacking in old-world primates, responsible for adding galactose-1,3 galactose to proteins.

Gammaretroviruses A species of retrovirus that includes PERV and MLV.

Genetically modified Refers to an organism whose genotype has been modified by application of biotechnology (e.g., gene transfer or chromosome set manipulation).

Genetic load The proportional amount by which the average fitness of a population is depressed for genetic reasons below that of the genotype with maximum fitness.

Genotype The genetic identity of an individual. Genotype often is evident by outward characteristics.

Germline cells Cells that contain inherited material that comes from the eggs and sperm, and that are passed on to offspring.

Hazard A substance or agent that, upon exposure, might result in a defined harm.

Helper, or packaging, cells Cells engineered to express retrovirus packaging proteins to produce retroviruses capable of infecting cells when the packaging genes have been deleted from the retrovirus genome. Such cells are widely used for the production of retrovirus vectors. The retroviruses produced are incapable of making progeny that can infect cells.

Homolog In diploid organisms, one member of a pair of matching chromosomes.

Homologous recombination Rearrangement of related DNA sequences on a different molecule by crossing over in a region of identical sequence.

Horizontal gene transfer Transmission of DNA between species, involving close contact between the donor's DNA and the recipient, uptake of DNA by the recipient, and stable incorporation of the DNA into the recipient's genome.

IgE (immunoglobulin E) A component of the human immune system implicated in the expression of allergies.

Insulators   Regions of DNA that separate (or insulate) the expression of one region from that of the next.

Integration   The covalent joining of a piece of DNA (like a provirus) into genomic DNA.

Intracytoplasmic sperm injection (ICSI)   Fertilization by direct injection of sperm into the cytoplasm of an egg.   ICSI can be used as a means of transfection.

Islets of Langerhans   The insulin-generating portion of the pancreas.

Knockin   Replacement of a gene by a mutant version of the same gene using homologous recombination.

Knockout   Inactivation of a gene by homologous recombination following transfection with a suitable DNA construct.

Lipofection   A method of transfection in which DNA is incorporated into lipid vesicles (liposomes), which then are fused to the membrane of the target cells.

Locus-control regions   Segments of DNA important for the correct and coordinated expression of large regions (such as those encoding hemoglobins).

Long terminal repeat (LTR)   A DNA sequence at the ends of the provirus (integrated DNA) of all retroviruses, derived during reverse transcription by duplication of sequences at the ends of the genome RNA. It contains most of the control elements necessary for expression of the provirus.

Mariner   A transposon originally isolated from insects, but related elements have been found in many animals, including humans.

Microinjection   The introduction of DNA into the nucleus of an oocyte, embryo, or other cell by injection through a very fine needle.

Mobilization   The transfer of genes from one place to another (in the same or a different cell or organism) mediated by a retrovirus or transposable element.

Murine leukemia virus (MLV) A retrovirus originally isolated from mice and widely used as the basis for retrovirus vectors.

Neomycin transferase A bacterial gene encoding resistance to several common antibiotics (kanamycin, neomycin, G418) widely used as a selectable marker in eukaryotic cells.

Niche An organism's place and function in the environment, defined by its utilization of resources.

Nuclear reprogramming Restoration of the correct embryonic pattern of gene expression in a nucleus derived from a somatic cell and introduced into an oocyte.

Nuclear transfer (NT) The generation of a new animal nearly identical to another one by injection of the nucleus from a cell of the donor animal into an enucleated oocyte of the recipient.

Phenotype The visible and/or measurable characteristics of an organism (i.e., how it appears outwardly) as opposed to its genotype, or genetic characteristics.

Plasmid A circular DNA molecule capable of replication in bacteria. Plasmids are the usual means of propagation of DNA for transfection or other purposes.

Pleiotropy A phenomenon whereby a particular gene affects multiple traits.

Polycations Large, positively-charged molecules often used to mediate transfection by reducing charge repulsion between DNA and the cell membrane.

Polygenic Refers to a trait or phenotype whose expression is the result of the interaction of numerous genes.

Porcine endogenous virus (PERV) Endogenous retrovirus of pigs closely related to MLV. Some PERVs can infect human cells.

Prion-related protein (PrP) A normal protein, expressed in the nervous system of animals, whose structure when altered (by interaction with altered copies of itself) is the cause of scrapie in sheep, BSE in cattle, and Creutzfeldt-Jakob disease in humans.

Promoter and enhancer insertion   Activation of expression of a gene by integration of a nearby provirus, bringing expression of the gene under the control of regulatory elements in the LTR of the provirus.

Promoter   A regulatory element that specifies the start site of transcription.

Pronuclear injection   The use of a fine needle to inject DNA into the nucleus of an unfertilized egg.

Provirus   The integrated DNA form of a retrovirus.

Recombinant   Refers to a genotype with a new combination of variable types, in contrast to parental type.

Recombinant DNA techniques   Procedures used to join together DNA segments in a cell-free system (an environment outside a cell or organism). Under appropriate conditions, a recombinant DNA molecule can enter a cell and replicate there, either autonomously or after it has become integrated into a cellular chromosome.

Retrovirus   An enveloped virus that replicates by reverse transcription of its RNA genome into DNA, followed by integration of the DNA into the cell genome to form a provirus. Expression of the provirus (as though it were a cellular gene) leads to the production of progeny virus particles.

Retrovirus vectors   Vector constructs in which the internal genes of a retrovirus are replaced by the gene of interest, flanked by the viral LTRs and packaging signals. After transfection of helper cells, the vector is packaged into virus particles.   Infection of target cells with these particles leads to integration of the gene into cellular DNA as part of a provirus.

Reverse transcription   The process of copying RNA into DNA, carried out by retroviral reverse transcriptase.

Risk   The likelihood of a defined hazard being realized, which is the product of two probabilities: the probability of exposure, $P(E)$, and the probability of the hazard resulting given that exposure has occurred, $P(H/E)$ (i.e., $R = P(E) \times P(H/E)$).

Scrapie   A disease, originally of sheep, but transmissible to other animals, characterized by neurological degeneration due to accumulation of a structural variant of PrP.

Sexual selection   The type of selection in which there is competition among males for mates and characteristics enhancing the reproductive success of the carrier are perpetuated irrespective of their survival value.

Selectable marker   A gene, usually encoding resistance to an antibiotic, added to a vector construct to allow easy selection of cells that contain the construct from the large majority of cells that do not.

Senescent cells   Animal cells that have nearly reached the limit of lifespan (usually around 50 doublings) in cell culture and are beginning to show signs of impending death.

Silencing   Shutdown of transcription of a gene, usually by methylation of C residues.

Sleeping beauty   A transposon related to *mariner*, originally isolated from fish.

Smoltification   The process through which a juvenile salmon becomes physiologically ready to enter salt water within its migratory life history.

Somatic cells   Cells of body tissues other than the germline.

Sperm-mediated transfection   Introduction of DNA into an oocyte by first mixing it with sperm, which then is used for fertilization.

Starlink   A brand of transgenic maize approved for animal feed only, but which also has been found in the human food supply.

T DNA   DNA encoded on a plasmid of *Agrobacterium* that integrates into the genome of a plant cell after being introduced into the cell by fusion.

Telomerase   The enzyme, absent from most somatic cells but present in germline cells, that restores telomeres to their normal length.

Telomeres   The simple repeated sequences at the ends of chromosomes that protect them from loss of coding sequence during replication. In the absence of telomerase, telomeres become progressively shorter with each cell division, and this shortening is the major cause of senescence of cells in culture.

Transfection Alteration of the genome of a cell by direct introduction of DNA, a small portion of which becomes covalently associated with the host cell DNA.

Transgene A gene construct introduced into an organism by human intervention.

Transposase The enzyme responsible for moving a transposon from one place to another.

Transposon A DNA element capable of moving (transposing) from one location in a genome to another in the same cell through the action of transposase.

Trojan gene A gene that drives a population to extinction during the process of spread as a result of destructive, self-reinforcing cycles of natural selection.

Vector A type of DNA, such as a plasmid or phage that is self-replicating and that can be used to transfer DNA segments among host cells. Also, an insect or other organism that provides a means of dispersal for a disease or parasite.

Vertical transmission Inheritance of a gene from parent to offspring.

Viremia Virus in blood.

Virion The extracellular form of a virus (i.e., a virus particle).

Xenotransplantation Transplantation of cells, tissues, or organs from one species to another.

Zoonotic infection Transmission of an infectious agent from an animal reservoir to humans.

Zygote A fertilized oocyte.

Appendix A

Workshop Agenda

Defining Science-based Concerns Associated with Products of

Animal Biotechnology: A Public Workshop

NRC Committee on Defining Science-based Concerns

Associated with Products of Animal Biotechnology

The National Academies

Green Building, Room 104
2001 Wisconsin Ave, NW
Washington, DC 20007

November 27, 2001

Agenda

OPEN SESSION

8:00–8:15 a.m.	Welcome and Opening Remarks
	John G. Vandenbergh and **Kim Waddell**
8:15–8:35 a.m.	How We Make "Them" and Where We're Headed
	Robert J. Wall, *USDA/ARS*
8:35–9:15 a.m.	Current Applications of Somatic Cell Cloning
	José B. Cibelli, *Advanced Cell Technology, Inc.*
9:15–9:35 a.m.	The Use of Transposons in Animal Biotechnology
	Perry B. Hackett, *Discovery Genomics/University of Minnesota*
9:35–10:00 a.m.	Discussion
10:00–10:15 a.m.	Break
10:15–10:35 a.m.	Food Animal Productivity and Welfare
	Paul B. Thompson, *Purdue University*
10:35–10:55 a.m	Defining Animal Biotechnology Policy: How Far Will Science and Regulation Be Able to Take Us?
	Jean Fruci, *Pew Initiative on Food and Biotechnology*

10:55–11:15 a.m.   A Framework for Identifying Hazards and Risks Associated with Transgenic Animals

Larisa Rudenko, *Integrative Biostrategies, L.L.C.*

11:15 a.m.–12:00 p.m.   Discussion

12:00–1:00 p.m.   Lunch/Break

1:00–1:40 p.m.   Cloning of Farm Animals: A Four-year Analysis

Michael D. Bishop, *Infigen, Inc.*

1:40–2:00 p.m.   European Perspectives on Animal Cloning and Biotechnology

Keith H. S. Campbell, *University of Nottingham*

2:00–2:20 p.m.   Food Allergenicity and Biotechnology

Samuel B. Lehrer, *Tulane University*

2:20–2:45 p.m.   Discussion

2:45–3:15 p.m.   Transgenic Insects: Potential Risk Assessment Issues

Marjorie A. Hoy, *University of Florida*

3:15–3:45 p.m.   Consumer Perspectives on Animal Biotechnology

Michael K. Hansen, *Consumer Policy Institute*

3:45–4:00 p.m.   Break

4:00–5:00 p.m.   Discussion

5:00 p.m.   Adjourn

Appendix B

Regulatory Framework for Animal Biotechnology

BACKGROUND

The regulatory framework for animal biotechnology consists of the agencies, statutes, regulations and policies under which: (1) standards are established for the care and treatment of animals used in biotechnology research and testing activities, (2) decisions are made about market access and conditions of use for the commercial products of animal biotechnology, and (3) government post-approval oversight is provided to verify that marketed products are in compliance with regulatory requirements.

The regulatory framework's standards and procedures for making market access decisions are particularly important to the study because they establish the general scope of the questions regulatory agencies will need to ask about the commercial products of animal biotechnology.

This discussion assumes that the initial products of animal biotechnology will involve: (1) modifications that affect the performance of the animal or attributes of products derived from the animal through the action of the expression product of an inserted gene; (2) animals modified to produce drugs, biologics, or other substances of commercial value; or (3) cloned animals.

The regulatory framework and the questions agencies will ask will vary depending the nature of the modification and the intended use of the resulting product(s).

AGENCIES

The Animal and Plant Health Inspection Service (APHIS) of USDA has jurisdiction over livestock used in biomedical research, teaching, or testing, to oversee compliance with the regulations for animal care and use promulgated by APHIS under the Animal Welfare Act (9 CFR Parts 1–4).

The Office of Laboratory Animal Welfare of the National Institutes of Health (NIH) has responsibility for the general administration and coordination of the Public Health Service Policy (1996) on the Humane Care and Use of Laboratory Animals, including livestock and poultry used in biomedically-related research activities.

The Center for Veterinary Medicine (CVM) in the Food and Drug Administration (FDA) has asserted primary jurisdiction over the first two categories of anticipated products of animal biotechnology noted in the background section above for purposes of making market access decisions, and setting conditions of use, but CVM's jurisdiction is not exclusive.

The Centers for Drug Evaluation and Research (CDER) and Biologics Evaluation and Research (CBER) in FDA also have jurisdiction over the products in category 2 to the extent they involve human drugs or biologics.

The Center for Veterinary Biologics in the USDA's Animal and Plant Health Inspection Service (APHIS) has jurisdiction over products in category 2 to the extent they are modified to produce animal biologics.

The Center for Food Safety and Applied Nutrition (CFSAN) in FDA has jurisdiction over milk, eggs and other edible products (other than meat and poultry products) to oversee compliance with limits on residues in the edible tissue and ensure the general wholesomeness and safety of the food.

The Food Safety and Inspection Service (FSIS) in USDA has jurisdiction over meat and poultry derived from genetically modified animals to oversee compliance with residue limits and ensure the general wholesomeness and safety of the food.

It is unclear whether any agency has jurisdiction to make market access decisions or establish conditions of use for cloned animals. FDA has said that it may have jurisdiction over human cloning through its new drug authority and that it expects human cloning experiments to be covered by investigational new drug applications (IND's). CVM has not taken a public position on its regulatory jurisdiction over animal cloning, but it is for now constrained not to take a position that is different in principle from the FDA position on human cloning. CFSAN and FSIS have jurisdiction to oversee the general wholesomeness and safety of the edible tissue of a cloned animal, just like any other animal.

STATUTES

The Animal Welfare Act (1966) provides APHIS with the authority to regulate warm-blooded animals used for biomedical research, teaching, and testing, with the exception of rats, mice, and birds. Agricultural animals used in agriculturally related (i.e. food and fiber) research, testing, or teaching, are specifically excluded by Congress from regulation under the Animal Welfare Act.

The Health Research Extension Act (1985), "Animals in Research", requires that guidelines be established for the proper care and treatment of vertebrate animals used in biomedical and behavioral research, and that assurance be provided that each applicant for a grant or contract from the National Institutes of Health or other national research institute complies with these guidelines. The Health Research Extension Act provided the statutory mandate for adoption of the Public Health Service Policy (see below). Under this Act, the NIH may suspend or revoke a grant or contract in the case of noncompliance.

The animal drug provisions of the Federal Food, Drug, and Cosmetic Act (FDCA) provide the basic authority for CVM's oversight of products in categories 1 and 2. FDA considers the expression product resulting from the genetic modification of an animal to be a "new animal drug" under section 201(v) of FDCA. New animal drugs must be licensed by FDA under section 512 of the FDCA based on a showing that the products are safe and effective for their intended use. "Safe" has reference to the safety of man or animal (section 201(u) of FDCA).

The general food safety provisions of the FDCA, the Federal Meat Inspection Act, and the Poultry Products Inspection Act provide CFSAN and FSIS authority to oversee the wholesomeness and safety of edible tissue from genetically modified animals.

The new drug provisions of FDCA (primarily section 505) and biologics provisions of the Public Health Service Act (primarily section 351) provide CDER and CBER authority to regulate (by requiring premarket scientific review and licensing) the safety and effectiveness of human drugs and biologics produced by genetically modified animals and to ensure that they are produced under conditions that ensure their purity and potency.

The Virus, Serums, and Toxins Act (VSTA) provides APHIS the authority to regulate the safety and effectiveness of animal biologics to ensure safety and effectiveness.

The National Environmental Policy Act (NEPA) requires all agencies to conduct an environmental assessment (EA) and, when there may be significant impact on the quality of the human environment, an environmental impact statement in connection with agency actions. Under NEPA, CVM would conduct an EA in connection with its approval of genetically modified animals under its animal drug licensing authority and seek measures to ameliorate any anticipated adverse environmental effects. NEPA does not override the market entry standards of the FDCA, and CVM is not legally empowered to deny approval of an animal drug based on its NEPA assessment. CVM asserts that its animal drug authority permits it to regulate the environmental impacts of genetically modified animals to the extent they adversely affect, directly or indirectly, the health of humans or animals. This presumably would include requiring mitigation actions and monitoring of environmental impacts. While NEPA is intended to provide for public consideration of the environmental impact of government actions, the FDCA's animal drug authorities and regulations make the licensing process confidential between the applicant and the agency and preclude disclosure of information contained in the new animal drug application (NADA) confidential until the product is approved.

REGULATIONS

APHIS has issued regulations under the Animal Welfare Act (9 CFR Part 1–4) setting forth procedures for the oversight of the use of mammals (except purpose-bred rats and mice) in biomedical research, teaching, and testing. The regulations require certain minimum standards of housing and care, and also stipulate that research, teaching, or testing activities involving animals first be approved by an Institutional Animal Care and Use Committee. Although the Animal Welfare Act potentially applies to all warm-blooded animals, APHIS has so far chosen not to regulate birds; hence poultry used in biomedical research are not regulated under the Act.

FDA has issued detailed regulations governing the animal drug approval process as it applies generally to animal drugs (21 CFR Parts 510–514). These were written prior to the advent of animal biotechnology. There are no FDA regulations specifically addressing animal biotechnology.

Similarly, FDA has issued procedural regulations for its human drug and biologics programs but no regulations specifically targeting products derived from genetically modified animals.

APHIS has issued procedural regulations governing its regulation of animal biologics under the Virus, Serums, and Toxins Act (9 CFR Parts 101–118).

FDA has issued regulations under NEPA, setting forth the procedures for EA's and EIS's (21 CFR Part 25).

POLICIES AND OTHER GUIDANCE

The Public Health Service (PHS) Policy on Humane Care and Use of Laboratory Animals (1986) describes the general standards and procedures to be used by institutions to comply with the provisions of the Health Research Extension Act, including approval of research activities involving animals by an Institutional Animal Care and Use Committee. The PHS policy also requires compliance with the Animal Welfare Act and the standards outlined in the Guide for the Care and Use of Laboratory Animals (1998) published by the Institute for Laboratory Animal Resources.

The Federation of Animal Sciences Societies has issued guidance for the oversight of food and fiber research and teaching using animals, the *Guide for the Care and Use of Agricultural Animals in Agricultural Research and Teaching* (the *Ag Guide*). Adoption of these guidelines by an institution is voluntary. The *Ag Guide* is recognized as a resource document by USDA-APHIS and in the Guide for the Care and Use of Laboratory Animals, and as a primary reference document by the Association for the Assessment and Accreditation of Laboratory Animal Care International (AAALACI), the major animal care accrediting body for U.S. research institutions.

CVM has issued informal guidance through articles and speeches on its general approach to overseeing animal biotechnology under its animal drug authority but has not issued any formal policy statements or guidance.

CBER has issued scientific guidance to the industry in the form of "Points to Consider in the Manufacture and Testing of Therapeutic Products for Human Use Derived From Transgenic Animals" (1995).

FSIS has issued a points-to-consider document on "Safety Evaluation of Transgenic Animals from Transgenic Animal Research."

About the Authors

JOHN G. VANDENBERGH, *Chair*, is Professor of Zoology at North Carolina State University, where he began teaching and managing a research program in behavioral endocrinology in 1990. He has published extensively on the behavior, genetics, and physiology of small mammals. His previous National Academies committee service includes the Committee on Understanding the Biology of Sex and Gender Differences, the Institute for Laboratory Animal Research (ILAR) committees on Care and Use of Laboratory Animals, Cost of and Payment for Animal Research, and the Revised Guide for the Care and Use of Laboratory Animals. He earned his Ph.D. in zoology from Pennsylvania State University.

ALWYNELLE (NELL) SELF AHL is USDA Fellow to the Center for the Integrated Study of Food, Animal, and Plant Systems at Tuskegee University. Dr. Ahl's interests include risk assessment for foodborne microbial pathogens, environmental epidemiology, the evolutionary biology of mammals, and the intersection of animals and humans in relation to public health. She received her Ph.D. in zoology and biochemistry from the University of Wyoming in 1967, and her D.V.M. from Michigan State University in 1987.

JOHN M. COFFIN is the American Cancer Society Research Professor of Molecular Biology and Microbiology at the Tufts University School of Medicine. He also is the Director of the HIV Drug Resistance Program at the National Cancer Institute in Frederick, Maryland. Coffin is recognized for his pioneering work in the use of genomic analysis to understand the biology of retroviruses, elucidating their genetic organization, mechanism of replication, recombination, and transduction. His areas of expertise include viruses, viral vectors, and transduction and their role in animal biotechnology.

WILLARD H. EYESTONE is a research associate professor in the Department of Large Animal Clinical Sciences at the Virginia-Maryland Regional College of Veterinary Medicine. He conducts basic and applied research in animal reproduction, with emphasis on embryo biotechnology. Eyestone was one of the original team members that produced the first transgenic cow as well as the first calves from *in vitro*-produced embryos. Dr. Eyestone received his Ph.D. from the University of Wisconsin in animal science.

ERIC M. HALLERMAN is an Associate Professor in the Department of Fisheries and Wildlife Sciences at Virginia Polytechnic Institute and State University. He also is a member of the Biotechnology Risk Assessment Steering Committee for the USDA. Dr. Hallerman's research interests include genetic improvement of aquaculture stocks, population genetics as applied to fisheries and wildlife management, genetics education, risk assessment/management and public policy regarding genetically modified fish and shellfish. He received his Ph.D. in fisheries and allied aquacultures from Auburn University in 1984.

TUNG-CHING LEE is Distinguished Professor of Food Science and Nutrition at Rutgers University. He is noted for his effective usage of basic science of chemistry and microbiology in studying applied programs in food technology. Dr. Lee's research interests include nutritional, safety, and toxicological aspects of food processing, food nutrification, aquaculture, feed technology, and biotechnologic applications in food science and technology. He received his Ph.D. in agricultural chemistry in 1970, from the University of California, Davis.

JOY A. MENCH is a professor in the Animal Science Department at the University of California, Davis. Her particular focus has been in the role of animal behavior, both social and individual, and the effects on animals of stress, crowding, handling, restraint, and other components of captivity. Dr. Mench has worked closely as a scientist and consultant with animal welfare and laboratory animal accreditation organizations over the past 15 years. She received her Ph.D. in neurobiology and ethology from the University of Sussex, England.

WILLIAM M. MUIR is Professor of Breeding and Genetics in the Department of Animal Sciences at Purdue University, as well as Director of the High Definition Genomics Center. Dr. Muir's research interests include the development of transgenic animals for enhancement of the environment and the profitability of farming, assessment of the environmental risk of transgenic organisms, and the linkage between quantitative and molecular genetics. Dr. Muir received his Ph.D. in population genetics from Purdue University in 1977.

R. MICHAEL ROBERTS is the Curator's Professor of Animal Science and Biochemistry at the University of Missouri. He is best known for his contributions in facilitating understanding of embryo-maternal communication during the early stages of pregnancy. He was the first to discover that early placentas produce interferons that mediate maternal recognition of the embryo in cattle and sheep. Dr. Roberts received his Ph.D. in plant physiology and biochemistry from Oxford University, England, in 1965.

THEODORE H. SCHETTLER is the Science Director for the Science and Environmental Health Network. He also is a Board Member for the Greater Boston Physicians for Social Responsibility (GBPSR). With GBPSR, Dr. Schettler serves as Co-Chair of the Committee on Human Health and the Environment. He holds a clinical appointment at Boston Medical Center and practices medicine at the East Boston Neighborhood Health Center. He holds an M.D. from Case-Western Reserve Medical School, and an M.P.H. from the Harvard School of Public Health.

LAWRENCE B. SCHOOK is Professor of Comparative Genomics in the Department of Animal Sciences and Veterinary Pathobiology at the University of Illinois. He also is Adjunct Professor of Veterinary Pathobiology at the University of Minnesota. Dr. Schook's professional interests include the development of genomic models to address animal health. His research laboratory has developed genetic markers and integrated maps, and has mapped economically important traits in livestock. He holds a Ph.D. in immunology from Wayne State University.

MICHAEL R. TAYLOR is Senior Fellow and Director of the Risk, Resource, and Environmental Management Division at Resources for the Future (RFF), a nonprofit natural resource and environmental research organization. He also leads a research program at RFF on policy and institutional issues affecting the success of the global food and agricultural system in the areas of food security in developing countries, food safety, and the natural resource and environmental sustainability of agriculture. Previously, Mr. Taylor served as Deputy Commissioner for Policy at the U.S. Food and Drug Administration and as Administrator of the Food Safety and Inspection Service of the U.S Department of Agriculture. Mr. Taylor received a J.D. in 1976 from the University of Virginia School of Law.

Board on Agriculture and Natural Resources Publications

Policy and Resources

Agricultural Biotechnology: Strategies for National Competitiveness (1987)
Agricultural Biotechnology and the Poor: Proceedings of an International Conference, (2000)
Agriculture and the Undergraduate: Proceedings (1992)
Agriculture's Role in K-12 Education: A Forum on the National Science Education Standards (1998)
Alternative Agriculture (1989)
Brucellosis in the Greater Yellowstone Area (1998)
Colleges of Agriculture at the Land Grant Universities: Public Service and Public Policy (1996)
Colleges of Agriculture at the Land Grant Universities: A Profile (1995)
Countering Agricultural Bioterrorism (2002)
Designing an Agricultural Genome Program (1998)
Designing Foods: Animal Product Options in the Marketplace (1988)
Ecological Monitoring of Genetically Modified Crops (2001)
Ecologically Based Pest Management: New Solutions for a New Century (1996)
Ensuring Safe Food: From Production to Consumption (1998)
Environmental Effects of Transgenic Plants: The Scope and Adequacy of Regulation (2002)

Emerging Animal Diseases: Global Markets, Global Safety: Workshop Summary, (2002)

Exploring Horizons for Domestic Animal Genomics. (2002)

Forested Landscapes in Perspective: Prospects and Opportunities for Sustainable Management of America's Nonfederal Forests (1997)

Future Role of Pesticides for U.S. Agriculture (2000)

Genetic Engineering of Plants: Agricultural Research Opportunities and Policy Concerns (1984)

Genetically Modified Pest-Protected Plants: Science and Regulation (2000)

Incorporating Science, Economics, and Sociology in Developing Sanitary and Phytosanitary Standards in International Trade (2000)

Investing in Research: A Proposal to Strengthen the Agricultural, Food, and Environmental System (1989)

Investing in the National Research Initiative: An Update of the Competitive Grants Program in the U.S. Department of Agriculture (1994)

Managing Global Genetic Resources: Agricultural Crop Issues and Policies (1993)

Managing Global Genetic Resources: Forest Trees (1991)

Managing Global Genetic Resources: Livestock (1993)

Managing Global Genetic Resources: The U.S. National Plant Germplasm System (1991)

National Capacity in Forestry Research (2002)

National Research Initiative: A Vital Competitive Grants Program in Food, Fiber, and Natural-Resources Research (2000)

New Directions for Biosciences Research in Agriculture: High-Reward Opportunities (1985)

Nutrient Requirements of Beef Cattle, Update (2000)

Nutrient Requirements of Dairy Cattle, Seventh Revised Edition (2001)

Nutrient Requirements of Nonhuman Primates, Second Revised Edition (2002)

Nutrient Requirements of Swine, Tenth Revised Edition (1998)

Pesticide Resistance: Strategies and Tactics for Management (1986)

Pesticides and Groundwater Quality: Issues and Problems in Four States (1986)

Pesticides in the Diets of Infants and Children (1993)

Precision Agriculture in the 21st Century: Geospatial and Information Technologies in Crop Management (1997)

Predicting Invasions of Nonindigenous Plants and Plant Pests (2002)

Professional Societies and Ecologically Based Pest Management (2000)

Publicly Funded Agricultural Research and the Changing Structure of U.S. Agriculture (2002)

Rangeland Health: New Methods to Classify, Inventory, and Monitor Rangelands (1994)

Regulating Pesticides in Food: The Delaney Paradox (1987)

Scientific Advances in Animal Nutrition: Promise for a New Century (2001)

Soil and Water Quality: An Agenda for Agriculture (1993)
Soil Conservation: Assessing the National Resources Inventory, Volume 1 (1986); Volume 2 (1986)
Sustainable Agriculture and the Environment in the Humid Tropics (1993)
Sustainable Agriculture Research and Education in the Field: A Proceedings (1991)
The Role of Chromium in Animal Nutrition (1997)
The Scientific Basis for Estimating Emissions from Animal Feeding Operations: Interim Report (2002)
The Scientific Basis for Predicting the Invasive Potential of Nonindigenous Plants and Plant Pests in the United States (2002)
The Use of Drugs in Food Animals: Benefits and Risks (1999)
Toward Sustainability: A Plan for Collaborative Research on Agriculture and Natural Resource Management (1991)
Understanding Agriculture: New Directions for Education (1988)
Use of Drugs in Food Animals: Benefits and Risks, The (1999)
Water Transfers in the West: Efficiency, Equity, and the Environment (1992)
Wood in Our Future: The Role of Life Cycle Analysis (1997)

Nutrient Requirements of Domestic Animals Series and Related Titles

Building a North American Feed Information System (1995)
Metabolic Modifiers: Effects on the Nutrient Requirements of Food-Producing Animals (1994)
Nutrient Requirements of Beef Cattle, Seventh Revised Edition, Update (2000)
Nutrient Requirements of Cats, Revised Edition (1986)
Nutrient Requirements of Dairy Cattle, Seventh Revised Edition (2001)
Nutrient Requirements of Dogs, Revised Edition (1985)
Nutrient Requirements of Fish (1993)
Nutrient Requirements of Horses, Fifth Revised Edition (1989)
Nutrient Requirements of Laboratory Animals, Fourth Revised Edition (1995)
Nutrient Requirements of Poultry, Ninth Revised Edition (1994)
Nutrient Requirements of Sheep, Sixth Revised Edition (1985)
Nutrient Requirements of Swine, Tenth Revised Edition (1998)
Predicting Feed Intake of Food-Producing Animals (1986)
Role of Chromium in Animal Nutrition (1997)
Scientific Advances in Animal Nutrition: Promise for the New Century (2001)
Vitamin Tolerance of Animals (1987)

Further information, additional titles (prior to 1984), and prices are available from the National Academies Press, 500 Fifth Street, NW, Washington, DC, 20001; 202–334–3313 (information only). To order any of the titles you see above, visit the National Academies Press bookstore at *http://www.nap.edu/bookstore.*

Index

Enzymes, 37, 69, 103

Ethical issues, 116, 118, 119-121
  animal health and welfare, 3, 4, 11-12,
  13-14, 22-23, 93, 104, 116, 118
  committee study at hand, methodology,
  2, 3, 4, 32
  conventional breeding *vs*
  biotechnology, 3, 104
  religious concerns, 23-24, 116, 118

European Union, steroids, 27

Exogenous pig viruses, 55-58

Exposure, 33, 75, 76-77
  human health, 6-7, 61

F

Farm animals, general, 4, 11, 12, 16, 17,
19, 23, 25, 30, 79
  see also specific species
  animal health and welfare, 93-94, 104
  historical perspectives, 20, 21, 22,
  23-24
  small-scale farms, 13

FDA, *see* Food and Drug Administration

Feces, 7, 62, 65

Federal government, 4, 112-115
  see also Food and Drug Administration;
  Legislation; Regulatory issues
  Army Corps of Engineers, 113
  Center for Food Safety and Applied
  Nutrition, 113-114
  Center for Veterinary Medicine, 114,
  114
  Department of Agriculture, 21, 26, 63,
  113
  Environmental Protection Agency, 113
  Fish and Wildlife Service, 113

Federal Food, Drug, and Cosmetic Act,
114

Feral species, effects on, 4, 10, 11, 33,
81-92, 113-115

  see also Aquatic organisms;
  Environmental concerns
  fish, 83, 84, 89-92, 109

Fertility and fertilization, 5, 9, 24-25
  animal health and welfare, 99
  artificial insemination (AI), 4, 9, 12. 19,
  23-24, 65, 77, 93-96
  germline modification, 34
  sterility, 29-30, 88

Fish, 7, 11, 16, 19, 21, 29-31, 39, 70-71
  allergenicity and hypersensitivity, 68
  animal health and welfare, 98-99
  feral species, 83, 84, 89-92, 109
  monosex fish, 30-31
  shellfish, 7, 11, 16, 21, 29-31, 39, 68,
  70-71, 92
  transfection, 36
  transposons, 37

Fish and Wildlife Service, 113

Food allergies, *see* Allergenicity and
hypersensitivity

Food and Drug Administration (FDA),
1-3 (passim), 23, 31, 63, 64, 111, 113-114
  bovine somatotropin, 28
  conventional breeding *vs*
  biotechnology, 3
  food safety, historical perspectives, 62,
  70-71
  xenotransplantation, 59

Food and food safety, 6-9, 16-17, 18-19,
61-72, 111, 117
  see also Beef and dairy cattle; Fish;
  Milk and milk products; Poultry and eggs
  age factors, 9, 65, 70
  allergenicity and hypersensitivity, 7-8,
  61-62, 66, 67, 68-69
  committee charge and methodology, 1,
  2, 6-9, 13, 61, 73
  embryos, 8, 24-26, 63-65
  ethical issues, 61, 63-65
  gene expression, 64-65, 66, 67, 69
  historical perspectives, 3, 5, 62, 70-71
  international perspectives, 62, 69
  labeling, 118